Canning & Preserving for Beginners

The Essential Canning Recipes & Canning Supplies Guide

Rockridge Press

ISBN: Print 978-1-62315-183-6 | eBook 978-1-62315-184-3

CONTENTS

AN INTRODUCTION TO CANNING

For generations home canning was common in many households and considered an integral part of feeding a family. Before the majority of households turned to supermarkets to supply their food, canning was a way to have enough food on hand during the winter months when gardens lay dormant and fresh fruits and vegetables weren't available. Farming families typically used canning to preserve a great deal of the family's food for the year. Most nonfarming families also commonly had a kitchen garden of some kind and used home canning to prepare and preserve what was not eaten fresh. Canning even remained a part of everyday life for many rural families long after the supermarket entered the scene.

Recently home canning has experienced an explosion of popularity, even among people who have never gardened or canned before. Many people have become interested in canning because they're concerned about economic instability. Rising food prices and unemployment rates have them worried about being able to feed their families well should times become hard. Canning can be hard work, but it's far more economical than purchasing fresh, frozen, or canned foods from the grocery store. By growing your own foods or purchasing them locally and in season when prices are at their lowest, you can avoid the heavy markups on out-of-season produce. By canning your own soups, stews, and other meals, you can realize huge savings on dining out or picking up takeout on those days when you don't have time to cook or forgot to thaw something for dinner.

Some people are uneasy about the quality and safety of commercially grown produce. They prefer to eat locally sourced foods, and they use home canning to preserve seasonal produce to eat throughout the year. Other people prefer to grow their own food, for either peace of mind or personal pleasure, and they want to be able to store their harvests to enjoy year-round. Talk to any canning enthusiast and you'll likely hear the person gush about the pleasure of looking at shelves laden with multicolored jars of wholesome food. You'll probably also hear about the peace of mind that canners get from knowing that no matter what might happen, they can reach into their pantry to put together a delicious and healthy meal.

Not only is home canning insurance against potential hard times and an extremely satisfying undertaking, it also can be great fun. It is most enjoyable, as well as efficient, when it's done as a group, with several hands pitching in to prep the ingredients, prepare the food, process the jars, and seal and label the day's work. The home canning process can be a time of togetherness, shared work, and shared bounty.

Whatever motivates you to begin canning your own food, you won't be disappointed in the rewards. This book has been created to help the most inexperienced beginner to can food safely and successfully using water bath canning, pressure canning, or both. You'll find plenty of the best-loved classic recipes for fruits, vegetables, condiments, sauces, soups, stews, and other canning favorites to get you started, and you'll learn how to use each canning method step by step.

Canning is a wonderful way to feed your family and live more sustainably and economically. This book takes you from the planning stages to that point when you, too, can look with pride at shelves lined with multicolored jars and say, "Wow! I did this!"

The Two Methods of Home Canning

There are two methods used in home canning: the water bath (or boiling water) method and pressure canning. The method to use depends on the type of food you will be canning.

The Water Bath (Boiling Water) Method

This method is often considered the best way for a beginner to start canning because the equipment is somewhat cheaper and the process is a little less involved. Water bath canning is used only for highly acidic foods such as tomatoes, berries, and pickles, which don't require longer processing times to discourage bacterial growth in the preserved foods. The vacuum-sealed acidic environment is enough to keep foods safe and delicious in storage.

With water bath canning, racks are placed in stockpots or water bath canners filled with water and the pot is brought to a boil on the stove top. The filled canning jars are placed on these racks and processed (i.e., boiled) for a specific amount of time according to the food being canned. Then the jars are cooled, which is when the lids seal shut. Once the jars have cooled and the seals are tight, rings are added to the jars and the food can be stored away.

The Pressure Canning Method

Pressure canning is done, as the name implies, in either a stove-top or electric pressure canner. These canners come in many sizes, and the one you choose depends primarily on how many jars you want to process at one time. As in water bath canning, metal racks are placed in the pressure canner, the canner is filled with water, and the jars are placed inside. The jars are processed under pressure (dictated by the foods you're canning) for a certain period of time (also dictated by the food you're canning) and then removed to cool and seal.

Pressure canning may seem intimidating to beginners, but it's easy to learn and incredibly rewarding because it allows you to safely can almost any food, including meats, stews, soups, chilies, and other meals.

You can decide which method you would like to start with. Your decision might be based on the equipment you have available, a friend or family member who can lead you through a particular method, or the types of foods that you're interested in canning. Whichever method you choose to learn first, you may want to borrow the equipment and purchase only a few jars. This way, you can limit your investment until you know for certain that canning is for you. Start out with very simple foods and recipes, such as dill pickles, tomato sauce, or canned peaches, and gradually work your way toward more complex recipes and even trying your own variations. When you're completely comfortable with both the science and the process of your first method of canning, then you can confidently start learning the other.

Since many people typically choose to start with the water bath method, this book begins with the water bath canning method and water bath recipes.

Note: You can actually process high-acid foods in a pressure canner, but water bath canning is quicker, so the choice is up to you.

Food Acidity and How It Affects Your Processing Method

Whether food should be processed in a water bath canner or a pressure canner depends on the acidity of the food you're canning. Acidity can be natural, as with many fruits, or it can be added. The term *pH* is a measure of the acidity in a food; the lower the pH value, the more acid in the food. You can increase the acidity level in foods by adding lemon juice, citric acid, or vinegar.

High-acid foods contain enough acid to block the growth of *Clostridium botulinum*. High-acid foods have a pH of 4.6 or lower. These include fruits; pickles; jams, jellies and marmalades; fruit butters; and sauerkraut.

Low-acid foods don't have enough acid to prevent the growth of the *Clostridium botulinum* bacteria, which is what causes botulism. Low-acid foods need to be heated to protect them. Low-acid foods have natural pH values higher than 4.6. These include meats, seafood, poultry, milk, and fresh vegetables, except for most varieties of tomatoes. When you create a mixture of low-acid and high-acid foods, that mixture also has a pH value above 4.6, unless the recipe also includes enough lemon juice, citric acid, or vinegar to make it a high-acid food. The chart below gives approximate pH levels for popular canning foods. Note that levels may vary with different varieties of the fruits and vegetables listed.

STRONG ACID				STRONG ALKALI		
1	2	3	4	5	6	7
	citrus		apples	carrots	corn	
		apricots		asparagus	peas	
	pickles		tomatoes	green beans	chicken	
		blackberries		beets	shrimp	
	gooseberries				fish	
		peaches			ham	
		plums	pears	ground beef		
		strawberries		potatoes	dairy products	
Use water bath canning method				Use pressure canning method		

Note: Although tomatoes are generally considered a high-acid food, some of them have pH values slightly above 4.6. You should either add lemon juice or citric acid to these varieties, or use a pressure canner to process them.

Botulinum bacteria are almost impossible to eliminate simply by bringing them to a boiling temperature (212 degrees F). This is why low-acid foods need to be processed at temperatures between 240 and 280 degrees F.

In a pressure canner, this means processing the food at 10 to 15 psi. The abbreviation *psi* means "pounds per square inch of pressure," which you measure by using the pressure gauge located on or near the lid of the canner. At temperatures of 240 to 280 degrees F, the bacteria in low-acid canned food can be destroyed in 20 to 100 minutes. The exact time it will take depends on the kind of food you're canning, the way it is packed, and the size of the jars you're using. The time needed to safely process low-acid foods in a water bath canner would range between 7 and 11 hours, but the time needed to process high-acid foods using a water bath canner only runs between 5 and 85 minutes.

Processing time using either pressure canning or the water bath method depends on:
- the type of food
- the altitude of your geographic location
- the recipe's acidity
- the size of the jars

Each of the recipes in this book includes processing times.

Fresh produce from your garden or a farmers' market is the best choice for canning.

Selecting Produce for Canning

Canning does not make mediocre foods taste more palatable, nor will it rescue food that's about to spoil. It's important to be choosy when selecting the foods you'll be canning; that way, your canned foods will be delicious and safe to eat.

Choose the Best Quality You Can Find

Fresh is best and fresh in season is even better. Canning began as a way to preserve the best flavors of a given season, in addition to putting food up for the non-growing season.

Pick fruits and vegetables that are at their very peak of flavor. This means buying in season as much as possible and locally, if available in your area. Foods picked before they're ripe and shipped thousands of miles will lack the flavor of locally sourced foods. This means you should be looking for strawberries in June and in the fall, fresh peas in the spring, and so on. Many types of produce have long growing seasons, especially if you live in a more temperate climate, so it's fairly easy to buy them fresh and flavorful for much of the year. Cucumbers, beans, and tomatoes are good examples.

If you're not able to grow your own produce, try to purchase foods for canning at a farmers' market or from a local farmer. Organic produce is always a good choice for canning, and farmers may be willing to give you a price break on the larger quantities for a day of canning.

Inspect Your Produce Carefully Before Processing

If you're buying your produce by the case, bushel, or bucket, be sure to carefully inspect each piece as you wash it. Discard anything that appears to have mold on it (either green or a thin, white coating near the stem). Also be sure to toss anything that has holes, cuts, or other openings that could be portals for bugs, insect eggs, or bacteria.

Tips to Avoid Spoilage

The last thing you want after carefully growing, choosing, and canning your wonderful foods is to lose some of them to spoilage. More important, you don't want anyone to become ill from eating spoiled food.

Canning is a safe way of preparing and preserving fresh foods as long as you follow the recipes and instructions carefully. The following additional tips can help you can foods successfully and minimize the chances of spoilage:

- Always follow the guidelines in the following sections for testing the seals on your canned goods before you store them away. If you are ever in doubt about the seal or if a jar has not sealed properly, either refrigerate it and use immediately, throw it away, or reprocess it. Reprocessing is recommended only for foods that do not contain meat, eggs, or seafood. (To reprocess, prepare another sterilized jar, bring the food to a boil in a saucepan, and pour it into the new jar. Cover and process for the required time, and allow it to sit undisturbed for another 12 to 24 hours.)

- Have enough clean towels on hand so you can wipe each jar with a fresh section of towel. Once a section of towel has become soiled with liquid or food, use a new section or a new towel.

- If you're canning more than one type of food, always wash your utensils, cutting boards, knives, and other tools before moving from one food to another.

- Can only what your family can eat or give away within a year. Always mark the jars accordingly, and eat the oldest foods first. A good practice is to write the canning date in marker on the lid and write the expiration date on the label. When adding new jars to a shelf, move the older jars to the front and store the newer jars behind them.

- Always store your canned goods in a cool, dark place where they will be safe from jostling. Never store canned foods in movable storage such as totes or under-bed storage boxes, as frequent moving can result in chipped or cracked jars and cause contamination.

SECTION 1

Water Bath Canning

1

WATER BATH CANNING:
SUPPLIES, INSTRUCTIONS, AND SAFETY

The equipment and supplies for water bath canning are easy to find and inexpensive. You don't need much, and you often can find used equipment or borrow some from a friend. If you're unsure whether canning is something you will do long term, stick with the essentials at first and add more supplies later on.

What You'll Need for Water Bath Canning

You'll need the following essential supplies to begin canning using the water bath method. The individual recipes will give you further guidance and any additional items needed. Use the checklist on the next page and "The Basic Instructions for Water Bath Canning" later in this chapter to help you get started.

- ☑ Large water bath canner
- ☑ One or more canning racks
- ☑ Thermometer for testing food and water temperatures
- ☑ Jar lifter (tongs used to lift hot jars from the canner)
- ☑ Jars (½ pint, pint, or quart sizes)
- ☑ Canning funnel
- ☑ New lids for each jar
- ☑ Bands for each jar
- ☑ Clean dish towels for wiping jar rims
- ☑ Clean dish towels for cooling the hot jars
- ☑ Butter knife or thin spatula for removing air bubbles
- ☑ Watch or kitchen timer
- ☑ Marker and labels

Choosing a Water Bath Canner

You can use a large aluminum or porcelain-covered steel stockpot for water bath canning, or you can purchase one made just for canning. The advantage of choosing a water bath canner is that its metal racks have handles for removing the jars from the pot. Some retailers even have beginner's sets that include the canner, a jar lifter, racks, and other utensils. If you choose to use a stockpot you already own, you can buy the racks and utensils separately from any canning supply vendor.

Water bath canners come in several sizes, and the label will tell you how many jars can be processed at once. Most water bath canners can process 14 to 18 (½ pint) jars, 7 to 8 pint jars, or 5 to 7 quart jars. For safe processing, the water level needs to be at least 1 inch above the tops of the jars; keep that in mind as you choose your canner and your jars.

A Note on Jars and Lids

It is perfectly acceptable to buy used canning jars. As long as the jars aren't cracked or chipped, they can be reused for many years. You can often find them at thrift shops, garage sales, and on sites like Freecycle or Craigslist. This can be a very economical way to go, especially if you begin canning in large quantities or you want to have several sizes of jars on hand. New jars are not expensive, generally costing ten dollars per dozen.

Although you may hear of people canning with recycled glass jars (from commercially packaged mayonnaise, applesauce, or other products), this is not recommended. These jars may not be made of glass that can withstand the high temperatures of canning, and lids and rings may not fit tightly enough on recycled jars. In any event, used canning jars can be had for pennies, so the cost savings from using recycled glass jars just isn't worth the risk.

You can also buy and safely use used rings (in fact, they often come with the used jars), but first carefully inspect them for warping, cracks, rust, and other damage. Rings are sold by the dozen or the case, and they are usually available for five dollars or less per dozen, so you may be better off discarding any used rings. You may want to reuse your own rings, however, once you've emptied the jars. Just be sure to inspect them before using them again.

Lids, however, *cannot* be reused. Once you break the seal to open your jars, the lids are likely warped and unable to reseal, even if they appear to be undamaged. Lids are also extremely inexpensive, so don't risk your health to save a few pennies.

The preceding list covers everything you really need to get started in water bath canning. If you find you love canning, you may want to add more equipment and utensils, and you'll have a better idea of what you do and do not really need once you gain some experience.

Basic Instructions for Water Bath Canning

1. Before you start preparing your food for canning, fill your canner halfway full with cold tap water. This will be the right water level for one load of pint jars, but you may need to add more water when processing quart jars or two or more racks of half-pint jars. You'll need 1 to 2 inches of water above the tops of the jars.

2. Prepare your ingredients while preheating the water in the canner to 140 degrees F for raw foods and 180 degrees F for hot-packed foods. (Check your recipe before beginning.)

3. Follow the recipe directions for preparing and packing your food. Use a butter knife or thin spatula to remove any air bubbles.

4. Wipe the rim of each jar with a clean cloth before centering the lid on each. Screw on the band, and adjust it until it is fingertip tight. Be sure to wipe the rims of the jars with a clean towel before placing the lids onto the jars. Fit the lids onto the jars, and lower them one by one with a jar lifter or by loading a rack and lowering it into the canner.

5. If you use a jar lifter to load the canner, be sure you situate the lifter below the screw band of the lid and keep the jars vertical at all times. If you tip the jars and the contents spill into the sealing area, not only might it leak into your canner but also it can prevent a safe seal.

6. If necessary, add enough water to reach 1 inch above the jars for short processing times and 2 inches for processing times of 30 minutes or more.

7. Turn the heat to high, place the lid on the canner, and bring to a vigorous boil.

8. Once the water is boiling vigorously, start your kitchen timer for the required processing time. **Processing time always starts from the boiling point.**

9. Keep the canner covered and the water at a hard boil throughout the processing time. If the water stops boiling at any point, bring it back to a boil and restart the timer.

10. When the timer goes off at the end of processing, turn off the heat and remove the canner lid. Wait 5 minutes before you begin removing the jars.

11. While you wait, lay one or more towels onto a flat, cleared surface. Use enough towels to allow you to have 1 inch of space between jars.

12. Lift the rack of jars out of the canner (or remove them one by one using the jar lifter), and place each one onto the towel(s). Let the jars sit undisturbed for 12 to 24 hours.

13. Check each lid to be sure that you have an airtight seal on each jar. There are three ways to check the seals:

   - *Option 1*: Press down on the middle of the lid with your finger or thumb. If the lid springs back up when you release your finger, the lid is not properly sealed.

   - *Option 2:* Tap the lid with a teaspoon or butter knife. A clear ringing sound indicates a good seal. If it makes a dull or thunking sound, the lid is not sealed.

   - *Option 3:* Hold the jar at eye level and look straight across the lid. The lid should be slightly concave (curved downward in the center). If the center of the lid is either slightly bulging or flat, it may not be sealed.

Note: If you find any jars that have not sealed properly, refrigerate them and use the food within a couple of days.

---

14. Secure the sealed lids with rings, and use a marker or canning label to list both the contents and the date of processing before storing the jars in a cool, dark place. Most canned foods can be safely eaten for a year after processing, but check with your agricultural extension or your canner's instruction guide for storage dates based on the type of food.

15. The altitude of your geographical location affects water bath canning processing times. See the next page for additional processing times needed for elevations over 1000 feet (305 meters). Also see the back of the book for the altitude of selected cities in the United States and Canada. To find the exact altitude of your location, use the search features on the EarthTools website (www.earthtools.org).

Water Bath Canning Altitude Chart

Altitude in Feet	Altitude in Meters	Additional Processing Time
1,000 – 3,000	305 – 914	5 Minutes
3,001 – 6,000	915 – 1,829	10 Minutes
6,001 – 8,000	1,830 – 2,438	15 Minutes
8,001 – 10,000	2,439 – 3,048	20 Minutes

Tips for Water Bath Canning Safety and Food Handling

- Always check the recipe, your manufacturer's guide, or your local agricultural extension office to be sure that the type of food you're canning has enough acidity to be canned in a water bath canner.

- If you're canning mixed foods containing both high- and low-acid foods (e.g., some soups, relishes, stews), they must be processed in a pressure canner to prevent bacterial growth and spoilage.

- Always wipe the rims of the jars with a clean towel before placing the lids on the jars. Food and liquids on the rims of the jars can prevent them from sealing properly.

- When selecting and preparing foods for canning, always check carefully for mold, bruises, holes, and cuts that may harbor bacteria or insect eggs.

- If children are present, be sure they're safely away from the stove and the canner during processing, and that they understand that jostling or moving the jars during cooling can interfere with proper and safe sealing.

- Always make sure your cooling jars are safe from jostling or tipping for at least 12 to 24 hours. You may want to keep pets and small children out of the kitchen during the cooling period.

2

PERFECT PICKLES

Pickles are a great way to start learning how to can with the water bath method. They're considered one of the easiest foods to master, and most people love pickles, so they're also very rewarding.

The Different Types of Homemade Pickles

There are several different types of homemade pickles:

- Quick-process or cold-pack pickles are perhaps the most common and simplest to prepare. These are either marinated in a cold vinegar solution for several hours or are packed in a hot vinegar solution, and then they are processed immediately. Most of the recipes included in this section are for quick-process pickles.

- Fermented pickles are generally soaked in a briny solution for 4 to 6 weeks before being packed and processed.

- Refrigerator pickles are never processed. They are marinated or soaked for about a week, packed into jars, and then stored in the refrigerator for up to 2 months.

- Fruit pickles, such as watermelon rind pickles or pickled peaches, are exactly what they sound like. They're usually cooked in a type of syrup solution before being packed and processed.

Tips for Perfect Pickles

Pickles are easy to make at home, but the following few tips can maximize your potential for the perfect pickle:

- Always select firm cucumbers (or other foods) for making pickles. Overripe or slightly rubbery cucumbers yield floppy, unappetizing pickles. Pickling does not crisp them up, so choose the freshest, firmest cucumbers you can find.

- If you're pickling fruits or other vegetables, such as tomatoes, it's best to choose produce that is completely ripe. Sweetness is not an issue in pickling, but texture is. Slightly under-ripe produce produces a crisper texture.

- Over-processing can turn crisp cucumbers into wilted, leathery pickles in just a couple of minutes, so be sure to double-check the recipe's recommended processing time, watch your timer, and turn off the heat promptly.

- Cucumber blossoms contain an enzyme that can make your pickles limp. It's best to cut off the blossom end (just a sliver) to ensure this isn't a problem.

- You can use any size cucumber for pickles, but you want something close to uniformity in each jar. This will ensure even processing and pickling. If your pickling cucumbers vary a bit in size, it can be helpful to separate them into one or two groups when you're getting ready. This way, you don't have jars waiting to be processed (or jars already being processed) while you pick through your cucumbers to find the appropriate sizes.

- Watch your water and vinegar measurements. Too much or too little (most often the case) can result in soft pickles.

- Always use pickling salt (found in the canning section of most supermarkets). Table salt has additives that can make the pickle juice cloudy and off-putting.

- If you find that your processed pickles are cloudy, a few things may be the cause. Spoilage is your first concern, so open the pickles and look and smell for any signs of spoilage. When in doubt, throw them out. Most often, cloudiness is caused by using the wrong pan. Always use a nonmetallic or coated metallic pan, as aluminum will react with the vinegar in the brine and cause cloudiness. This cloudiness doesn't affect safety, but it is unappetizing.

- You can substitute fresh dill and dill seed one for the other. For every quart of pickles, use 3 heads of fresh dill or 1 to 2 tablespoons of dill seed.

- Generally, burpless cucumbers are not good for pickling. They contain an enzyme that softens the pickles during the fermentation process.

Classic Dill Pickles

Almost everyone loves dill pickles, and they're almost essential for a great sandwich or burger. These dill pickle slices are a cinch to make and a real crowd pleaser.

- 4 cups apple cider vinegar
- 4 cups water
- ¾ cup granulated sugar
- ½ cup pickling salt
- 3 tablespoons pickling spice*
- 5 whole bay leaves

- 5 large garlic cloves
- 2½ teaspoons mustard seed
- 5 heads fresh dill
- 13½ cups pickling cucumbers
- 5 pint jars, lids, and bands

* not the packets sold in the canning section, but the mixture sold in the spice section

1. Fill the canner with enough water to cover the jars. Boil the water, reduce the heat to low, place the jars in the water, and simmer until ready to use.1

2. Trim the blossom ends of the cucumbers, and cut into ¼-inch slices.

3. Tie the pickling spices up in a small piece of cheesecloth, and secure with kitchen twine.

4. Combine the vinegar, water, sugar, pickling salt, and pickling spice packet in a large stainless steel or porcelain-coated saucepan, and bring them to a boil over medium-high heat. Stir well to dissolve the salt and sugar.

5. Reduce the heat to medium, and simmer for 15 minutes or until the spices have infused the pickling liquid.

6. Place 1 bay leaf, 1 garlic clove, ½ teaspoon mustard seed, and 1 head of dill into each of the jars.

7. Pack the sliced cucumbers into the hot jars, leaving about ½ inch of headspace.

8. Ladle enough hot pickling liquid into each jar to cover the cucumbers, yet still leave ½ inch of headspace.

9. Remove air bubbles, wipe the rims, center the lids, and screw on the bands and adjust until they are fingertip tight.

10. Place the jars in the canner and bring to a boil. Make sure there is at least 1 inch of water covering the jars.

11. Process for 15 minutes, adjusting for altitude. Remove the jars from the canner and cool.

Sweet Gherkins

Sweet gherkins are a hit with everyone, especially children. If you can find them, it's fun to make these with the really tiny cucumbers. Otherwise, just choose the smallest pickling cucumbers you can find.

- 4 pounds cucumbers (3 to 4 inches long)
- 4 cups granulated sugar
- 3¾ cups white vinegar
- 3 tablespoons pickling salt
- 4 teaspoons celery seed
- 4 teaspoons ground turmeric
- 1½ teaspoons mustard seed
- 10 half-pint jars, lids, and bands

1. Trim the blossom ends of the cucumbers, and cut into quarters lengthwise.

2. Cover the cucumbers with boiling water, and let them stand for 2 hours.

3. Fill the canner with enough water to cover the jars. Boil the water, reduce the heat to low, place the jars in the water, and simmer until ready to use.

4. Drain the cucumbers and pack vertically into hot jars, leaving ¼ inch of headspace.

5. Combine the remaining ingredients in a nonmetallic or porcelain-coated pot, and bring to a boil.

6. Pour the hot liquid over each jar of cucumbers, leaving ¼ inch of headspace.

7. Remove air bubbles, wipe the rims, center the lids, and screw on the bands and adjust until they are fingertip tight.

8. Place the jars in the canner and bring to a boil. Make sure there is at least 1 inch of water covering the jars.

9. Process for 10 minutes, adjusting for altitude. Remove the jars from the canner and cool.

Bread and Butter Pickles

Bread and butter pickles are an old-fashioned favorite that combines the sweetness of a sweet pickle with an extra bit of tang. They're wonderful as is or added to potato, chicken, or egg salad.

- 4 pounds cucumbers
- 8 small onions, sliced
- ½ cup pickling salt
- 5 cups granulated sugar
- 4 cups white vinegar

- 2 tablespoons mustard seed
- 2 teaspoons celery seed
- 1½ teaspoons ground turmeric
- ½ teaspoon ground cloves
- 7 pint jars, lids, and bands

1. Trim the blossom ends of the cucumbers, and cut into quarters lengthwise.

2. In a large plastic tub or container, combine the cucumbers, onions, and pickling salt. Cover with crushed ice and mix well. Set aside for 3 hours. Drain the cucumbers, rinse, and drain again.

3. In a large nonmetallic or porcelain-coated pot, combine the sugar, vinegar, mustard and celery seeds, turmeric, and cloves. Bring to a boil, add the cucumber mixture, and return to a boil. Remove from the heat.

4. Fill the canner with enough water to cover the jars. Boil the water, reduce the heat to low, place the jars in the water, and simmer until ready to use.

5. Carefully ladle the hot mixture into the pint jars, leaving ½ inch of headspace.

6. Remove air bubbles, wipe the rims, center the lids, and screw on the bands and adjust until they are fingertip tight.

7. Place the jars in the canner and bring to a boil. Make sure there is at least 1 inch of water covering the jars.

8. Process for 15 minutes, adjusting for altitude. Remove the jars from the canner and cool.

Refrigerator Kosher Dills

These pickles are very similar to the ones you find in the refrigerated section of the grocery store. They're crisp, refreshing, and absolutely delicious. This is a refrigerator pickle, so no processing is needed.

- 6 pounds pickling cucumbers (4 inches long)
- 40 sprigs fresh dill
- 2 large red onions, thinly sliced
- 8 large cloves garlic, sliced
- 1 quart water
- 1 quart white vinegar
- ¾ cup granulated sugar
- ½ cup pickling salt
- 4 quart jars, lids, and bands

1. Trim the blossom ends of the cucumbers, and cut into quarters lengthwise.

2. In a very large bowl (or two large bowls), combine the cucumbers, fresh dill, red onions, and garlic and set aside.

3. In a Dutch oven, combine the water, white vinegar, sugar, and pickling salt. Bring it to a boil and simmer, stirring just until all of the sugar and salt is dissolved.

4. Pour the liquid over the cucumber mixture and allow to cool to room temperature, stirring often to make sure all of the cucumbers marinate evenly.

5. Place the cucumbers into the quart jars vertically, fully cover with the liquid, secure with lids and bands, and place in the refrigerator. Allow the pickles to sit for at least 24 hours in the refrigerator before eating, although they'll taste even better over the next few days. These will stay fresh in the refrigerator for up to 2 months.

Watermelon Rind Pickles

If you've never tried watermelon rind pickles, you might think they sound odd. But the texture of the skin and the white portion of the flesh of the watermelon are remarkably similar to cucumbers. This recipe will also work beautifully with honeydew or cantaloupe rind, winter squash, and even pumpkin! Change the fruit with the season, and give the pickles away as gifts.

- 4 pounds watermelon rind (about half of a medium watermelon)
- 2 cups white vinegar
- 1 quart plus 2 cups water, divided
- ¼ cup pickling salt (or more if needed)
- 4 cups granulated sugar
- 3 cinnamon sticks
- 1 teaspoon whole cloves
- 1 teaspoon whole allspice
- 1 lemon, thinly sliced
- 6 pint jars, lids, and bands

1. Trim all of the red flesh from the watermelon rind, and then cut the rind into 1-inch-thick slices. This will make it easier for you to peel the tough outer skin away with a vegetable peeler or paring knife. Once peeled, cut the rind into chunks about 1-inch square.

2. Place the rind in a very large bowl, and combine a quart of water and ¼ cup pickling salt, mixing well. Pour over the rind. (If this is not enough to cover the rind, add another quart of water and another ¼ cup pickling salt.)

3. Cover and refrigerate overnight or for 12 hours. Drain, rinse, and drain again.

4. Fill the canner with enough water to cover the jars. Boil the water, reduce the heat to low, place the jars in the water, and simmer until ready to use.

5. In a large saucepot, combine the 2 cups of water with the sugar, cinnamon sticks, cloves, allspice, and lemon slices. Bring to a boil.

6. Add the watermelon rind a handful at a time until all have been added to the boiling liquid. Reduce the heat to medium low, and simmer for 30 minutes or until the rind is clear.

7. Pack into the hot jars, and distribute the boiling liquid evenly among them.

8. Remove air bubbles, wipe the rims, center the lids, and screw on the bands and adjust until they are fingertip tight.

9. Place the jars in the canner and bring to a boil. Make sure there is at least 1 inch of water covering the jars.

10. Process for 15 minutes, adjusting for altitude. Remove the jars from the canner and cool.

Pickled Green Beans

Pickled green beans are absolutely addictive. They're easy to eat, crunchy, and packed with flavor. Try a jar of these instead of a bag of potato chips! This is also a great recipe to use on asparagus; just blanch the asparagus for 1 minute before beginning.

- 4 pounds green (string) beans
- 5 teaspoons crushed red pepper flakes
- 5 teaspoons mustard seed
- 5 teaspoons dill seed
- 10 large cloves garlic, slightly crushed
- 5 cups white vinegar
- 5 cups water
- ½ cup pickling salt
- 10 pint jars, lids, and bands

1. Fill the canner with enough water to cover the jars. Boil the water, reduce the heat to low, place the jars in the water, and simmer until ready to use.

2. Trim the ends and any strings from the green beans. Rinse and drain.

3. Into each jar, put ½ teaspoon each of the red pepper flakes, mustard seed, and dill seed. Add one clove of garlic to each jar.

4. Pack the green beans vertically into each jar until fairly tightly packed.

5. In a saucepan, combine the vinegar, water, and pickling salt and bring to a boil, stirring until the sugar and salt are dissolved.

6. Pour enough of the vinegar mixture into each pint jar, leaving ½ inch of headspace.

7. Remove air bubbles, wipe the rims, center the lids, and screw on the bands and adjust until they are fingertip tight.

8. Place the jars in the canner and bring to a boil. Make sure there is at least 1 inch of water covering the jars.

9. Process for 5 minutes, adjusting for altitude. Remove the jars from the canner and cool.

Pickled Banana Peppers

If you can keep everyone from snacking on these straight out of the jar, they make a wonderful topping for hot dogs, burgers, and sub sandwiches. They're also great to use on steaks, as a pizza topping, or in salads. This recipe works equally well for bell peppers. Use red, orange, yellow, or a combination of colors to get the best flavor, as green bell peppers aren't quite sweet enough.

- 1 pound banana peppers
- 4 cups white vinegar
- 1¼ cups granulated sugar
- 1 teaspoon mustard seed
- 1 teaspoon celery seed
- 4 half-pint jars, lids, and bands

1. Remove the stems, cores, and some or all of the seeds from the peppers, and slice into ½-inch rings.

2. Place the pepper rings into a small bowl filled with ice and water to crisp them up.

3. Fill the canner with enough water to cover the jars. Boil the water, reduce the heat to low, place the jars in the water, and simmer until ready to use.

4. In a saucepan, bring the vinegar, sugar, mustard seed, and celery seed to a rolling boil.

5. Divide the peppers evenly between the jars.

6. Pour enough of the hot pickling juice into the jars to leave ½ inch of headspace.

7. Remove air bubbles, wipe the rims, center the lids, and screw on the bands and adjust until they are fingertip tight.

8. Place the jars in the canner and bring to a boil. Make sure there is at least 1 inch of water covering the jars.

9. Process for 10 minutes, adjusting for altitude. Remove the jars from the canner and cool.

Pickled Jalapeño Peppers

Try this recipe for making your own jarred jalapeño peppers, just like the ones you buy in the store. If you love cooking Tex-Mex or Mexican food, your own home-canned peppers will quickly become a staple item in your pantry. Note: Most of the heat of the peppers is in the seeds. If you like them mild, remove all of the seeds before processing. If you like your peppers really hot, leave most of the seeds in. If you're somewhere in the middle, play with the amount of seeds until you get just the right level of heat.

- 2 quarts jalapeño peppers
- 2 cups white vinegar
- 2 cups water
- 1 teaspoon pickling salt
- 4 pint jars, lids, and bands

1. Remove the stems, cores, and some or all of the seeds from the peppers, and then slice into ½-inch rings. If you prefer to leave the peppers whole, cut a small slit into each of two sides of the peppers to keep them from exploding during processing.

2. Fill the canner with enough water to cover the jars. Boil the water, reduce the heat to low, place the jars in the water, and simmer until ready to use.

3. Combine the white vinegar and water and heat just to a simmer.

4. Pack the peppers tightly into the jars, and pour the hot liquid over the peppers, leaving ½ inch of headspace. Add ¼ teaspoon of salt to each jar.

5. Remove air bubbles, wipe the rims, center the lids, and screw on the bands and adjust until they are fingertip tight.

6. Place the jars in the canner and bring to a boil. Make sure there is at least 1 inch of water covering the jars.

7. Process for 15 minutes, adjusting for altitude. Remove the jars from the canner and cool.

Pickled Beets

Pickled beets are a sweet treat that was once standard fare at picnics and the dinner table. Although they're not quite as popular as they once were, they're incredibly tasty and well worth canning. Note: You'll need to wear disposable latex gloves while canning beets, as they'll stain your fingers purple for days!

- 6 quarts of fresh beets
- 1 teaspoon whole cloves
- 1 teaspoon whole allspice
- 2 cinnamon sticks

- 2 cups granulated sugar
- 2 cups white vinegar
- 2 cups water
- 8 pint jars, lids, and bands

1. Wash the beets, place them in a large pot of boiling water, and boil for 20 minutes. Drain and allow them to cool until they can be handled.

2. Trim off the ends of the beets and then peel the skin from the beets. They should slip off quite easily. Cut larger beets in quarters and leave small beets whole.

3. Fill the canner with enough water to cover the jars. Boil the water, reduce the heat to low, place the jars in the water, and simmer until ready to use.

4. Combine the cloves, allspice, cinnamon sticks, sugar, and vinegar with the 2 cups of water, and bring to a boil in a large stockpot. Stir to dissolve the sugar and then add the beets.

5. Cover, reduce the heat to medium, and simmer for 10 minutes. Remove the cinnamon sticks.

6. Pack the beets into the hot jars, and pour enough liquid into each to cover the beets and leave ½ inch of headspace.

7. Remove air bubbles, wipe the rims, center the lids, and screw on the bands and adjust until they are fingertip tight.

8. Place the jars in the canner and bring to a boil. Make sure there is at least 1 inch of water covering the jars.

9. Process for 30 minutes, adjusting for altitude. Remove the jars from the canner and cool.

Pickled beets, combined with fresh carrots and onions, make a tangy, flavorful side dish.

Sweet Refrigerator Pickled Onion Rings

If you're an onion lover, you'll really enjoy this preparation for home-canned pickled onion rings. These are great for topping sandwiches, burgers, and hot dogs. This is a refrigerator pickle, so no processing needed. They cannot be stored at room temperature, but they'll keep for two months in the refrigerator.

- 4 large white onions
- 2 cups apple cider vinegar
- 2 cups water
- ¼ cup granulated sugar
- 2 teaspoons pickling salt
- 4 pint jars, lids, and bands

1. Fill the canner with enough water to cover the jars. Boil the water, reduce the heat to low, place the jars in the water, and simmer until ready to use.

2. Peel the onions and cut into thin slices. Separate the slices into rings and divide between the jars.

3. In a saucepan, mix together the vinegar, water, sugar, and pickling salt and bring to a boil. Pour over the onion rings, leaving ½ inch of headspace.

4. Cover tightly and refrigerate for at least 48 hours before using. These will keep well in the refrigerator up to 2 months.

Italian-Style Pickled Vegetables

This is a great recipe to use when you have just a few vegetables waiting to be used. Serve it as a topping for sandwiches or as an antipasto. You can use banana peppers in place of the red peppers and red onions in place of yellow if you like.

- 4 pounds pickling cucumbers
- 2 pounds sliced white onions
- 4 cups celery, cut in 1-inch pieces
- 2 cups sliced carrots
- 2 cups 1-inch sweet red pepper pieces
- 2 cups bite-sized cauliflower florets
- 5 cups white vinegar

- ¼ cup yellow mustard
- ½ cup pickling salt
- 3½ cups granulated sugar
- 3 tablespoons celery seed
- 2 tablespoons mustard seed
- ½ teaspoon whole cloves
- 10 pint jars, lids, and bands

1. Trim the blossom ends of the cucumbers, and cut into ¼-inch slices.

2. In a large bowl, combine the cucumbers, onions, celery, carrots, peppers, and cauliflower and cover with 2 cups of ice and enough cold water to cover. Let stand for 1 hour to crisp up the veggies.

3. Fill the canner with enough water to cover the jars. Boil the water, reduce the heat to low, place the jars in the water, and simmer until ready to use.

4. In a mixing bowl, combine the vinegar, mustard, pickling salt, sugar, celery and mustard seeds, and cloves and mix well. Pour into a large saucepan and bring to a boil.

5. Drain the vegetable mixture and add to the saucepan of pickling liquid. Cover and return to a boil.

6. Drain vegetables, reserving the pickling liquid.

7. Pack the vegetables into the pint jars, leaving 1 inch of headspace. Pour the pickling liquid over the veggies, being sure to cover while leaving ½ inch of headspace.

8. Remove air bubbles, wipe the rims, center the lids, and screw on the bands and adjust until they are fingertip tight.

9. Place the jars in the canner and bring to a boil. Make sure there is at least 1 inch of water covering the jars.

10. Process for 5 minutes, adjusting for altitude. Remove the jars from the canner and cool.

3

JAMS, JELLIES, AND PRESERVES

Homemade jams, jellies, and preserves are some of the favorite types of foods to can. They're inexpensive to make and are usually made in small batches to achieve the right consistency, so you can make them with a small amount of fruit in a small block of time. Home-canned jams, jellies, and preserves are also delicious; they make wonderful and much-appreciated gifts for holidays, birthdays, housewarmings, and other occasions.

Introduction to Home-Canned Jams, Jellies, and Preserves

Before you get started on canning your own jams, jellies, and preserves, it's important to know the difference between the three.

- Jam is made with crushed fruit and its juice. Most people like to leave it fairly chunky, although you can puree it or make it smoother if you and your family prefer it that way. Fruit butters are very similar to jams, although the fruit is usually pureed after it has been cooked down, to create a very smooth texture.

- Jelly is made with the juice of the fruit rather than the fruit itself, and once it jells up, it is relatively clear. It's spreadable, but not as smooth as jam.

- Preserves are made with whole fruit or fruit cut into pieces. Preserves are quite chunky, but still spreadable, and you get more of the fruit's texture than you would with jam.

Jam, jellies, and preserves are cooked prior to canning, so they do take a little more time than many pickles, but the flavor makes the extra time well worthwhile.

Using Pectin

Many recipes for jams, jellies, and preserves use added pectin to help them thicken properly. Pectin is a naturally occurring chemical in fruits, but some fruits have more than others. Apples have quite a bit of pectin, especially in the core and peel, while strawberries have very little. Commercial pectin is made by extracting the pectin from apples or from the peel of citrus fruits, and it comes in powdered or liquid form. Liquid pectin is sometimes preferred for more delicate or less sweet fruits. Widely available, it is usually found in the canning section of most supermarkets.

Jams, jellies, and preserves can be made without pectin if enough sugar is used or if the fruit you're using has enough pectin in it. Using pectin, however, lets you use less sugar and also shortens the cooking time by quite a bit. Many recipes require only about 1 minute of cooking time to reach the proper thickness. Using pectin is very similar to using cornstarch to thicken a gravy or sauce.

Tips for Making Jams, Jellies, and Preserves

Here are some useful tips to making your jams, jellies, and preserves delicious and successful:

- Making these in small batches helps ensure that the fruit will cook quickly, and the color and flavor will be better.

- When you're creating your own jam and jelly recipes, remember this guideline: For every cup of fruit you use, you'll want to add ¾ cup of sugar. For example, 4 cups of fruit will require 3 cups of sugar. Some low-sugar recipes and recipes using apples and other high-pectin fruits can use less, but this is a good general rule.

- If you are using very ripe or especially sweet fruit, add 1 to 2 tablespoons of fresh or bottled lemon juice. The acid in the lemon juice will help the mixture thicken.

- You do need to watch out for scorching or burning fruit, as the sugar in the recipe can burn quickly. The best ways to prevent this are to watch your heat, adjusting it to a lower setting if needed, and to stir almost constantly during the cooking process.

- To test your jam or preserves to see if they're done, take a spoonful out of the pan and set it aside. If it holds its shape after about a minute, you should be ready to start canning the mixture.

Strawberry Jam

Strawberry jam may be the quintessential home-canned fruit spread. Strawberries ripen in June and in the fall, so buy large quantities during their peak seasons. Either make jam and can it right away, or wash, dry, and trim them and then freeze them in resealable bags until you're ready to can. This recipe works equally well for blueberries, boysenberries, and elderberries. Just adjust the sugar according to the sweetness of the fruit.

- 5 cups hulled strawberries
- ¼ cup fresh or bottled lemon juice
- 6 tablespoons liquid pectin
- 7 cups granulated sugar
- 8 half-pint jars, lids, and bands

1. Fill the canner with enough water to cover the jars. Boil the water, reduce the heat to low, place the jars in the water, and simmer until ready to use.

2. Using the low speed on a blender or food processor, pulse the strawberries once or twice so you're left with a chunky mixture. Or place the strawberries in a gallon-sized resealable bag, and pound a few times with the bottom of a small saucepan.

3. Combine the strawberries and lemon juice in a large saucepan, and slowly stir in the pectin, mixing well. Bring to a full rolling boil over high heat, stirring constantly.

4. Add the sugar to the liquid, stirring until completely dissolved. Return the mixture to a full boil. Boil for 1 minute, stirring constantly.

5. Remove the pan from the heat. If there is any foam on the surface (harmless but unattractive), skim it with a small strainer or a spoon.

6. Ladle the hot jam into the jars, leaving ¼ inch of headspace.

7. Remove air bubbles, wipe the rims, center the lids, and screw on the bands and adjust until they are fingertip tight.

8. Place the jars in the canner and bring to a boil. Make sure there is at least 1 inch of water covering the jars.

9. Process for 10 minutes, adjusting for altitude. Remove the jars from the canner and cool.

Variations:

Try these yummy variations on strawberry jam: To make Balsamic Strawberry Jam, reduce the lemon juice to 1 tablespoon and add 3 tablespoons of balsamic vinegar. This makes the jam a little tangier and gives it a nice zing. To make Strawberry-Lime Jam, add the zest of one large lime to the strawberries during cooking.

Raspberry Jam

Raspberries are a delectable fruit, and raspberry jam is one of the most popular flavors. Look for seedless varieties of berries to make things easier for yourself. Seeded varieties need to be pressed through a sieve or processed through a food mill to remove the seeds. This recipe works equally well for blackberries and black raspberries.

- 3½ cups (about 5 pints) crushed raspberries
- ¼ cup lemon juice

- 7 cups granulated sugar
- 3-ounce packet liquid pectin
- 8 half-pint jars, lids, and bands

---

1. Fill the canner with enough water to cover the jars. Boil the water, reduce the heat to low, place the jars in the water, and simmer until ready to use.

2. Using the low speed on a blender or food processor, pulse the raspberries once or twice so you're left with a chunky mixture. Or place the raspberries in a gallon-sized resealable bag, and pound a few times with the bottom of a small saucepan.

3. Combine the raspberries, lemon juice, and sugar in a large saucepan. Bring the mixture to a full rolling boil over high heat, stirring frequently.

4. Stir in the pectin all at once and continue to boil for 1 minute, stirring constantly.

5. Remove the pan from the heat, and skim off any foam.

6. Ladle the hot jam into the canning jars, leaving ¼ inch of headspace.

7. Remove air bubbles, wipe the rims, center the lids, and screw on the bands and adjust until they are fingertip tight.

8. Place the jars in the canner and bring to a boil. Make sure there is at least 1 inch of water covering the jars.

9. Process for 10 minutes, adjusting for altitude. Remove the jars from the canner and cool.

Blackberry Preserves

Blackberries are coveted as one of the finest treats of summer. They make a beautiful jar of preserves, and canning them is one of the best ways to enjoy their delicious flavor beyond their short growing season. This recipe works equally well with black raspberries.

- 3 pounds fresh blackberries
- 1¾-ounce package powdered pectin
- ¼ cup lemon juice
- 3 cups granulated sugar
- 12 half-pint jars, lids, and bands

1. Rinse the blackberries and pat dry. If using seeded berries, press through a sieve or food mill to remove the seeds.

2. Fill the canner with enough water to cover the jars. Boil the water, reduce the heat to low, place the jars in the water, and simmer until ready to use.

3. Combine the blackberries, pectin, and lemon juice in a large stockpot and bring to a boil.

4. Cook for 1 minute, stirring constantly. Add the sugar, stirring well, and return to a boil. Boil for 1 minute, stirring constantly.

5. Remove the pan from the heat, and let the mixture stand for 5 minutes, stirring every 30 seconds.

6. Ladle the hot mixture into the canning jars, leaving ¼ inch of headspace.

7. Remove air bubbles, wipe the rims, center the lids, and screw on the bands and adjust until they are fingertip tight.

8. Place the jars in the canner and bring to a boil. Make sure there is at least 1 inch of water covering the jars.

9. Process for 10 minutes, adjusting for altitude. Remove the jars from the canner and cool.

Variations:

To make Blackberry-Lime Preserves, add the zest of two limes to the berries during cooking. For Almond-Kissed Blackberry Preserves, add 1 teaspoon of almond extract after removing the berries from the heat.

Orange Marmalade

Orange marmalade sounds like a difficult recipe to make, but it's really quite easy. This makes a beautiful presentation, so it's a great home-canned product to give away as a gift.

- 2¼ pounds sliced oranges, peels on
- Zest and juice of 1 lemon
- 6 cups water
- 9 cups granulated sugar
- 8 half-pint jars, lids, and bands

1. Fill the canner with enough water to cover the jars. Boil the water, reduce the heat to low, place the jars in the water, and simmer until ready to use.

2. Combine the oranges, lemon zest, lemon juice, and water in a large saucepan. Bring to a boil over high heat, stirring constantly.

3. Reduce the heat and boil gently, stirring occasionally, for 40 minutes. Place the lid loosely (leaving ¼ inch of space) on the pot, and continue cooking for another 30 minutes, stirring occasionally.

4. Return the orange mixture to a boil over medium-high heat, stirring constantly. Once the water is boiling, stir in the sugar. Boil, stirring occasionally, until the mixture thickens nicely, about 15 minutes.

5. Remove the pan from the heat, and test the consistency by removing a spoonful and allowing it to sit. If after 1 minute the thickness is what you would like it to be, skim off any foam.

6. Ladle the hot mixture into the canning jars, leaving ¼ inch of headspace.

7. Remove air bubbles, wipe the rims, center the lids, and screw on the bands and adjust until they are fingertip tight.

8. Place the jars in the canner and bring to a boil. Make sure there is at least 1 inch of water covering the jars.

9. Process for 10 minutes, adjusting for altitude. Remove the jars from the canner and cool.

Variations:

You can also make wonderful marmalade from tangerines or kumquats. If using kum-quats, you will likely need to add a bit more sugar to taste. Grapefruit and lemon or lime marmalades are also delicious. Again, you will probably need to add more sugar until the marmalade reaches a sweetness level that you like.

Peach Freezer Jam

Peaches are at their juicy best when bought from local growers at the height of the season, usually early summer. Eat as many as you can fresh from the tree, but make this wonderful jam to enjoy their flavor long after summer has passed. Note: Ball and Kerr both make plastic canning containers specifically for freezing. They're great for both freezer jams and for freezing small quantities of soups, stews, and other meals.

- 2 pounds fresh peaches
- 2 tablespoons lemon juice
- 4½ cups granulated sugar
- ¾ cup water
- 6 tablespoons powdered pectin
- 8 half-pint jars, lids, and bands

1. Bring a large pot of water to a boil, and use a metal colander to drop the whole peaches in. Boil for 30 seconds. Place the colander into a sink or large bowl of ice water to stop the cooking process.

2. Slip the peels from the peaches, cut in half, remove the pits, and chop the peaches finely.

3. Combine the chopped peaches and the lemon juice in a large bowl. Add the sugar, mixing well. Let stand for 10 minutes.

4. Fill the canner with enough water to cover the jars. Boil the water, reduce the heat to low, place the jars in the water, and simmer until ready to use.

5. In a small saucepan, combine the water and pectin. Bring to a boil over high heat, and boil hard for 1 minute, stirring constantly.

6. Remove from heat, and add the pectin mixture to the peach mixture. Let it stand for 3 minutes.

7. Ladle the mixture into the canning jars, leaving ½ inch of headspace.

8. Remove air bubbles, wipe the rims, center the lids, and screw on the bands and adjust until they are fingertip tight.

9. Place the jars in the canner and bring to a boil. Make sure there is at least 1 inch of water covering the jars.

10. Remove the jars from the canner and cool.

11. Let the jam stand in the refrigerator until it is set, but no more than 24 hours. Serve immediately or freeze for up to 1 year. Once frozen jam is opened, it can be kept in the refrigerator for up to 3 weeks.

Mixed Berry Jam

This simple recipe makes the best of summer berries by combining them into one delicious jam. Once you've learned to make it, feel free to play with the ratios. For instance, you can make it half-and-half with blueberries and blackberries, or leave out the blueberries altogether and substitute extra blackberries and raspberries.

- 1 cup crushed blueberries
- 1 cup crushed strawberries
- 1 cup crushed blackberries
- 1 cup crushed raspberries

- 4½ tablespoons powdered pectin
- 3 cups granulated sugar
- 8 half-pint jars, lids, and bands

1. Fill the canner with enough water to cover the jars. Boil the water, reduce the heat to low, place the jars in the water, and simmer until ready to use.

2. Combine all of the berries in a large saucepan. Slowly stir in the pectin.

3. Bring the berry mixture to a full rolling boil over high heat, stirring constantly.

4. Add the sugar, stirring to dissolve completely. Return the mixture to a full boil. Boil hard for 1 minute, stirring constantly.

5. Remove the pan from the heat and skim off any foam.

6. Ladle the hot mixture into the canning jars, leaving ¼ inch of headspace.

7. Remove air bubbles, wipe the rims, center the lids, and screw on the bands and adjust until they are fingertip tight.

8. Place the jars in the canner and bring to a boil. Make sure there is at least 1 inch of water covering the jars.

9. Process for 10 minutes, adjusting for altitude. Remove the jars from the canner and cool.

Easy-Peasy Grape Jelly

Grape jelly is a favorite with kids and is easy to make once you get the hang of it. This recipe makes it especially easy by using bottled grape juice. It's fun to try different types of grape juice for making jelly. Experiment with white or red grape juice as well as Concord grape.

- 2 cups unsweetened grape juice
- 3½ cups granulated sugar
- 3-ounce packet liquid pectin
- 4 half-pint jars, lids, and bands

1. Fill the canner with enough water to cover the jars. Boil the water, reduce the heat to low, place the jars in the water, and simmer until ready to use.

2. Combine the grape juice and sugar in a large saucepan, and bring to a boil, stirring constantly.

3. Stir in the pectin and boil for 1 minute longer, stirring constantly.

4. Remove the pan from the heat, and skim off any foam.

5. Quickly pour the hot jelly mixture into the canning jars, leaving ½ inch of headspace.

6. Remove air bubbles, wipe the rims, center the lids, and screw on the bands and adjust until they are fingertip tight.

7. Place the jars in the canner and bring to a boil. Make sure there is at least 1 inch of water covering the jars.

8. Process for 5 minutes, adjusting for altitude. Remove the jars from the canner and cool.

Apricot Jam

Apricot Jam

Apricots are luscious when bought ripe and in season. Making a jam of them lets you enjoy them at their peak flavor all year round.

• 3½ cups chopped apricots (about 30 apricots), skin on	• 5¾ cups granulated sugar
• ¼ cup lemon juice	• 3-ounce packet liquid pectin
	• 6 half-pint jars, lids, and bands

1. Fill the canner with enough water to cover the jars. Boil the water, reduce the heat to low, place the jars in the water, and simmer until ready to use.

2. Combine the apricots, lemon juice, and sugar in a large saucepan. Bring the apricot mixture to a full rolling boil over high heat, stirring frequently.

3. Quickly stir in the pectin, squeezing the entire contents from the package. Continue to boil the mixture for 1 minute, stirring constantly.

4. Remove the pan from the heat, and skim off any foam.

5. Ladle the hot jam into the canning jars, leaving ¼ inch of headspace.

6. Remove air bubbles, wipe the rims, center the lids, and screw on the bands and adjust until they are fingertip tight.

7. Place the jars in the canner and bring to a boil. Make sure there is at least 1 inch of water covering the jars.

8. Process for 10 minutes, adjusting for altitude. Remove the jars from the canner and cool.

Variations:

For a nice change of flavor, try Apricot-Almond Jam. Add 1 teaspoon of almond extract to the apricots after removing the pan from the heat.

Apple Butter

Apple butter was used for generations (along with dried apples, applesauce, and cider) as a way of preserving a huge apple harvest. Homemade apple butter is absolutely delicious spread on pancakes, waffles, muffins, or toast. This recipe substitutes a slow cooker for the hours of standing and stirring that making apple butter used to require. Note: Apple varieties that cook down well are best for apple butter. Try Red Delicious, Yellow Delicious, Rome, or any combination of these apples.

- 3 cups granulated sugar
- 2 teaspoons ground cinnamon
- ½ teaspoon ground cloves
- ½ teaspoon ground nutmeg
- ¼ teaspoon salt
- 5 pounds apples, peeled and thinly sliced
- 6 half-pint jars, lids, and bands

1. Combine the sugar, cinnamon, cloves, nutmeg, and salt. Mix well.

2. Place the apples into the slow cooker in layers, alternating with the sugar mixture. As you add each layer of apples, press down to pack them in tightly.

3. Cover and cook on high for 1 hour. Reduce the heat to low. Remove the lid and stir well, breaking up the apples.

4. Return the cover to the slow cooker, and cook for 8 more hours or until the mixture is thick and caramel brown, stirring occasionally.

5. Reduce the heat to low and uncover. Cook for 1 hour more.

6. Fill the canner with enough water to cover the jars. Boil the water, reduce the heat to low, place the jars in the water, and simmer until ready to use.

7. Ladle the hot mixture into the canning jars, leaving ½ inch of headspace.

8. Remove air bubbles, wipe the rims, center the lids, and screw on the bands and adjust until they are fingertip tight.

9. Place the jars in the canner and bring to a boil. Make sure there is at least 1 inch of water covering the jars.

10. Process for 10 minutes, adjusting for altitude. Remove the jars from the canner and cool.

Cherry Preserves

If you have a cherry pitter (or the patience to remove the stones yourself), by all means make these preserves with fresh fruit, especially if you have your own trees. If not, this recipe uses thawed frozen cherries, which come already pitted.

- 6 cups (around 2 pounds) cherries, pitted
- 3½-ounce powdered pectin

- 3¼ cups granulated sugar, divided
- ½ teaspoon pure vanilla extract
- 6 half-pint jars, lids, and bands

---

1. Fill the canner with enough water to cover the jars. Boil the water, reduce the heat to low, place the jars in the water, and simmer until ready to use.

2. Place the cherries in a large, heavy saucepan.

3. Combine the pectin with ¼ cup of the sugar, and then stir into the cherries.

4. Bring to a full boil over high heat, stirring frequently.

5. Add the remaining 3 cups of sugar and return to a boil, stirring constantly, for 1 more minute.

6. Remove the pan from the heat, and skim off any foam.

7. Immediately ladle the hot preserves into the canning jars, leaving ¼ inch of headspace.

8. Remove air bubbles, wipe the rims, center the lids, and screw on the bands and adjust until they are fingertip tight.

9. Place the jars in the canner and bring to a boil. Make sure there is at least 1 inch of water covering the jars.

10. Process for 10 minutes, adjusting for altitude. Remove the jars from the canner and cool.

4

SALSAS AND RELISHES

Making your own salsas and relishes is both simple and fun. It's easy to exercise your creativity and customize flavors to your family's liking once you get the hang of making your own.

When making your own tomato salsas and relishes, be sure to use produce that is at its peak of ripeness. Processing will not improve the flavor of under-ripe tomatoes.

With fruit salsas such as mango or peach, you can use fruit that is just slightly under-ripe, as you want the fruit to hold its shape somewhat, and you don't want the salsa to be too sweet.

Sweet Pickle Relish

Making your own sweet pickle relish is simple and fun. Most of the time needed will be spent on chopping up your ingredients, but because the flavor is so much better than store-bought relish, you won't mind a bit.

- 4 cups diced cucumbers, with peel
- 1½ cups diced red bell pepper
- 2 cups diced red onion
- 1 cup diced celery
- ¼ cup salt

- 3½ cups granulated sugar
- 2 cups white vinegar
- 1 teaspoon celery seed
- 1 teaspoon mustard seed
- 6 pint jars, lids, and bands

---

1. Combine all vegetables in a large bowl, sprinkle with salt, add water to cover, and let stand for 5 hours. Drain well and press out remaining liquid.

2. Fill the canner with enough water to cover the jars. Boil the water, reduce the heat to low, place the jars in the water, and simmer until ready to use.

3. Combine the remaining ingredients in a saucepan and bring to a boil, stirring until the sugar is completely dissolved.

4. Stir in the diced vegetables, reduce the heat to medium, and simmer for 10 minutes, stirring occasionally.

5. Pack into canning jars, leaving ½ inch of headspace.

6. Remove air bubbles, wipe the rims, center the lids, and screw on the bands and adjust until they are fingertip tight.

7. Place the jars in the canner and bring to a boil. Make sure there is at least 1 inch of water covering the jars.

8. Process for 10 minutes, adjusting for altitude. Remove the jars from the canner and cool.

Onion Relish

Onion relish has a similar taste to sweet pickle relish and is a great way to use an abundance of fresh onions. It's especially delicious made with Vidalia onions but can also be prepared using yellow or red onions. Try it on hot dogs or added to dips, stews, and chilies.

- 5 pounds sweet onions (such as Vidalia or Walla Walla)
- ¼ cup salt
- 1 pint apple cider vinegar
- 1 teaspoon ground turmeric
- 4 ounces chopped pimiento
- 1 cup granulated sugar
- 1 teaspoon pickling spices
- 8 half-pint jars, lids, and bands

1. Chop the onions very fine, either by hand or in a food processor.

2. Place the onions into a large glass bowl or shallow casserole dish, sprinkle the salt over the onions, and allow to rest in the refrigerator for 1 hour.

3. Fill the canner with enough water to cover the jars. Boil the water, reduce the heat to low, place the jars in the water, and simmer until ready to use.

4. Drain the liquid from the onions. Combine the onions in a large pot with the vinegar, turmeric, pimiento, and sugar. Tie the pickling spices into a small square of cheesecloth and add to the pot.

5. Bring the onion mixture to a boil, and allow them to cook until the onions are transparent.

6. Pack the onions with the cooking liquid into the canning jars, leaving ½ inch of headspace.

7. Remove air bubbles, wipe the rims, center the lids, and screw on the bands and adjust until they are fingertip tight.

8. Place the jars in the canner and bring to a boil. Make sure there is at least 1 inch of water covering the jars.

9. Process for 10 minutes, adjusting for altitude. Remove the jars from the canner and cool.

Corn Relish

Corn Relish

Corn relish is an old-fashioned favorite that is still widely enjoyed in the South and Southwest. It's a wonderful side dish for Tex-Mex or Mexican cooking and also works very well as a side for barbecues and picnics. It's very colorful and makes a wonderful gift.

- 22 ears fresh, raw corn
- 1 cup diced red bell pepper
- 1 cup diced green bell pepper
- 1¼ cups diced celery
- ¾ cup diced sweet onion
- 1½ cups granulated sugar
- 2½ cups white vinegar
- 2 cups water
- 1½ teaspoons mustard seed
- 1 teaspoon celery seed
- ½ teaspoon ground turmeric
- 1 teaspoon salt
- 6 pint jars, lids, and bands

1. In a very large stockpot or two smaller pots, boil the corn in well-salted water for 4 minutes, working in batches if necessary.

2. As you remove the ears from the boiling water, place them in a large bowl (or bowls) of ice and water to stop the cooking process.

3. Drain the corn well and cut from the cob using a corn cutter or a sharp knife. You should end up with about 10 cups of cut corn.

4. In a large pot, combine the corn, peppers, celery, onion, sugar, vinegar, water, mustard and celery seeds, turmeric, and salt, stirring well to combine. Simmer over medium heat for 20 minutes, stirring occasionally.

5. Fill the canner with enough water to cover the jars. Boil the water, reduce the heat to low, place the jars in the water, and simmer until ready to use.

6. Pack the corn relish into the jars, leaving ½ inch of headspace.

7. Remove air bubbles, wipe the rims, center the lids, and screw on the bands and adjust until they are fingertip tight.

8. Place the jars in the canner and bring to a boil. Make sure there is at least 1 inch of water covering the jars.

9. Process for 15 minutes, adjusting for altitude. Remove the jars from the canner and cool.

Chow-Chow

Chow-chow is another old-fashioned recipe that is still very popular in Amish communities and the Deep South. Also known as piccalilli, gardeners love it as a way to use up an abundance of green tomatoes before the cool weather hits.

- 5 cups chopped green tomatoes
- 5 cups chopped cabbage
- 2 cups chopped red bell pepper
- 1½ cups chopped sweet onion
- ¼ cup pickling salt
- 2½ cups apple cider vinegar

- 1 cup packed light brown sugar
- 1 tablespoon mustard seed
- 2 large cloves garlic, finely chopped
- 1 teaspoon celery seed
- ½ teaspoon red pepper flakes (optional)
- 4 pint jars, lids, and bands

1. Combine all of the chopped vegetables in a large glass or plastic bowl. Add the pickling salt and stir well until completely combined. Cover and refrigerate for 4 hours or overnight.

2. Drain the vegetables and rinse thoroughly; drain again.

3. Fill the canner with enough water to cover the jars. Boil the water, reduce the heat to low, place the jars in the water, and simmer until ready to use.

4. In a large stockpot, combine the vinegar, brown sugar, mustard seed, garlic, celery seed, and red pepper flakes and bring to a boil. Reduce the heat to low and continue to simmer for 5 minutes.

5. Add the drained vegetables and bring back to a boil. Reduce the heat to low once more and simmer for 10 minutes.

6. Use a slotted spoon to pack the vegetables into the canning jars. Add enough pickling liquid to each jar to cover, while leaving ¼ inch of headspace.

7. Remove air bubbles, wipe the rims, center the lids, and screw on the bands and adjust until they are fingertip tight.

8. Place the jars in the canner and bring to a boil. Make sure there is at least 1 inch of water covering the jars.

9. Process for 10 minutes, adjusting for altitude. Remove the jars from the canner and cool.

Mild Salsa

Salsa is a wonderful dish that can easily be customized for the degree of heat your family enjoys. It's also a fun recipe for using your creativity, as there are so many types of salsas you can make according to what you like and have on hand. This is a great basic, mild recipe.

- 12 pounds cored and quartered tomatoes
- 4 green bell peppers, chopped
- 3 large yellow onions, chopped
- 1 red bell pepper, chopped
- 1 stalk celery, chopped
- 15 large cloves garlic, chopped
- 4–5 jalapeño peppers, seeded and chopped
- 2 (12-ounce) cans tomato paste
- 1¾ cups white vinegar
- ½ cup granulated sugar
- ¼ cup pickling salt
- ¼ –½ teaspoon hot pepper sauce
- 10 pint jars, lids, and bands

1. Place all of the tomatoes in a large stockpot, and cook over medium heat for 20 minutes. Drain the tomatoes, reserving 2 cups of the liquid, and return just the tomatoes to the pot.

2. Stir in the green bell peppers, onions, red bell pepper, celery, garlic, jalapeños, tomato paste, vinegar, sugar, pickling salt, hot pepper sauce, and reserved tomato liquid. Bring the mixture to a boil and reduce the heat to medium-low.

3. Simmer uncovered for 1 hour, stirring frequently.

4. Fill the canner with enough water to cover the jars. Boil the water, reduce the heat to low, place the jars in the water, and simmer until ready to use.

5. Ladle the hot mixture into the canning jars, leaving ¼ inch of headspace.

6. Remove air bubbles, wipe the rims, center the lids, and screw on the bands and adjust until they are fingertip tight.

7. Place the jars in the canner and bring to a boil. Make sure there is at least 1 inch of water covering the jars.

8. Process for 20 minutes, adjusting for altitude. Remove the jars from the canner and cool.

Variations:

For some fun changes in the flavor of your salsa, make Tex-Mex Salsa by adding 1 teaspoon ground cumin and 1 teaspoon chopped fresh cilantro to the pot before cooking. You can also make Rainbow Salsa by substituting 2 pounds peaches for the tomatoes and adding them into the pot with the peppers and other fresh veggies.

Salsa Verde

Salsa verde has a fresh, clean, but spicy flavor. It's most commonly made with tomatillos, but green tomatoes work just as well and are often more readily available, especially if you have a garden. Great as a dip for chips, salsa verde is also wonderful in white bean or chicken chilies and on tacos.

- 7 cups (about 12 medium-sized) green tomatoes
- 5 –7 jalapeño or favorite hot peppers, seeded and chopped
- 2 cups chopped red onion
- 2 large cloves garlic, finely chopped
- ½ cup lime juice
- ½ cup chopped fresh cilantro
- 2 teaspoons ground cumin
- 1 teaspoon oregano
- 1 teaspoon pickling salt
- 1 teaspoon freshly ground black pepper
- 6 half-pint jars, lids, and bands

1. Fill a large stockpot with water and bring to a boil. Reduce the heat to low, add the tomatoes, and simmer for 5 minutes to loosen the skin. Immediately plunge the tomatoes into a large bowl of ice-cold water to stop the cooking process. Then peel, core, and quarter the tomatoes.

2. Fill the canner with enough water to cover the jars. Boil the water, reduce the heat to low, place the jars in the water, and simmer until ready to use.

3. Combine the tomatoes, hot peppers, onion, garlic, and lime juice in a large saucepan over high heat and bring to a boil.

4. Stir in the cilantro, cumin, oregano, pickling salt, and black pepper. Reduce the heat to medium heat and simmer for 5 minutes, stirring occasionally.

5. Ladle the hot salsa into the canning jars, leaving ½ inch of headspace.

6. Remove air bubbles, wipe the rims, center the lids, and screw on the bands and adjust until they are fingertip tight.

7. Place the jars in the canner and bring to a boil. Make sure there is at least 1 inch of water covering the jars.

8. Process for 20 minutes, adjusting for altitude. Remove the jars from the canner and cool.

Mango Salsa

Mango salsa is a real treat that delivers all the spice of regular salsa, with just a hint of sweetness. Choose not-quite-ripe mangoes, as they'll flavor the salsa without making it too sweet.

- 6 cups (3 to 4 large-sized) diced mango
- 1½ cups diced red bell pepper
- ½ cup chopped sweet onion, such as Vidalia or Walla Walla
- ½ teaspoon crushed red pepper flakes
- 2 teaspoons chopped garlic
- 2 teaspoons chopped fresh ginger
- 1 cup packed light brown sugar
- 1¼ cups apple cider vinegar
- ½ cup water
- 6 half-pint jars, lids, and bands

1. Fill the canner with enough water to cover the jars. Boil the water, reduce the heat to low, place the jars in the water, and simmer until ready to use.

2. Combine all of the ingredients into a large stockpot. Bring to a boil over high heat and stir until the sugar is completely dissolved. Reduce the heat to medium-low and simmer for 5 minutes.

3. Use a slotted spoon to pack the vegetable mixture into the hot half-pint jars, adding enough hot liquid to cover, while leaving ½ inch of headspace.

4. Remove air bubbles, wipe the rims, center the lids, and screw on the bands and adjust until they are fingertip tight.

5. Place the jars in the canner and bring to a boil. Make sure there is at least 1 inch of water covering the jars.

6. Process for 10 minutes, adjusting for altitude. Remove the jars from the canner and cool.

Pineapple Salsa

Pineapple salsa is one of the sweeter fruit salsas and is incredibly addictive. You'll probably have trouble saving any of it for anything other than scooping with chips, but if you can reserve a jar, this is wonderful on hot dogs, mixed into ground meat for Polynesian-style burgers, or used as a topping for baked chicken or fish.

- 6 cups (about 3 medium-sized) cored and chopped pineapple
- 1 cup lemon juice
- ½ cup lime juice
- ½ cup pineapple juice
- ½ cup chopped and seeded Anaheim chili peppers or hot banana peppers
- 2 tablespoons chopped green onion
- 2 tablespoons chopped fresh cilantro
- 2 tablespoons packed brown sugar
- 6 half-pint jars, lids, and bands

1. Combine all the ingredients in a large stainless stockpot. Stir well to combine completely.

2. Bring to a boil over medium-high heat, stirring constantly. Reduce the heat to medium and simmer, stirring frequently, for 10 minutes.

3. Fill the canner with enough water to cover the jars. Boil the water, reduce the heat to low, place the jars in the water, and simmer until ready to use.

4. Ladle the hot salsa into the canning jars, leaving ½ inch of headspace.

5. Remove air bubbles, wipe the rims, center the lids, and screw on the bands and adjust until they are fingertip tight.

6. Place the jars in the canner and bring to a boil. Make sure there is at least 1 inch of water covering the jars.

7. Process for 15 minutes, adjusting for altitude. Remove the jars from the canner and cool.

Hot and Spicy Salsa

For more adventurous palates, this zesty salsa can be as fiery as you like. This recipe is fairly hot, but you can easily adjust the amount of peppers and hot pepper sauce to your liking.

- 4 cups peeled, cored, and chopped tomatoes
- 2 cups chopped, seeded green chili peppers
- 2 cups chopped, seeded jalapeño peppers
- ¼ cup chopped yellow onions
- 4 large cloves garlic, finely chopped
- 2 cups apple cider vinegar
- 1 teaspoon ground cumin
- 1 tablespoon chopped fresh oregano
- 1 tablespoon chopped fresh cilantro
- 1½ teaspoons pickling salt
- 4 pint-sized jars, lids, and bands

1. Combine all of the ingredients in a large saucepan, and bring the mixture to a boil over medium-high heat, stirring frequently. Reduce the heat and simmer for 20 minutes, stirring occasionally.

2. Fill the canner with enough water to cover the jars. Boil the water, reduce the heat to low, place the jars in the water, and simmer until ready to use.

3. Ladle the hot salsa into the canning jars, leaving ½ inch of headspace.

4. Remove air bubbles, wipe the rims, center the lids, and screw on the bands and adjust until they are fingertip tight.

5. Place the jars in the canner and bring to a boil. Make sure there is at least 1 inch of water covering the jars.

6. Process for 15 minutes, adjusting for altitude. Remove the jars from the canner and cool.

Mango Chutney

There are as many ways to prepare chutney as there are ways to enjoy it. Common in Indian cooking, chutney is a wonderful condiment for steaks, chops, fish, and stews. This recipe is our version of a classic chutney recipe, adapted for home canning.

- 4 pounds (about 10 large-sized) unripe mangoes
- 4½ cups granulated sugar
- 3 cups white vinegar
- 2½ cups chopped yellow onion
- 2½ tablespoons grated fresh ginger
- 1½ tablespoons chopped fresh garlic
- 2½ cups golden raisins
- 4 teaspoons chili powder
- 1½ teaspoons pickling salt
- 6 pint jars, lids, and bands

1. Peel the mangoes and chop by hand. Using a food processor will puree them too much.

2. Mix the sugar and vinegar in a large stockpot. Bring to a boil over medium-high heat and boil for 5 minutes, stirring frequently.

3. Add the onion, ginger, garlic, raisins, chili powder, and pickling salt to the pot and stir well to combine. Add in the chopped mango and stir again.

4. Bring the mixture back to a boil and then reduce the heat and simmer for 25 minutes, stirring occasionally.

5. Fill the canner with enough water to cover the jars. Boil the water, reduce the heat to low, place the jars in the water, and simmer until ready to use.

6. Ladle the hot chutney into the canning jars, leaving ½ inch of headspace.

7. Remove air bubbles, wipe the rims, center the lids, and screw on the bands and adjust until they are fingertip tight.

8. Place the jars in the canner and bring to a boil. Make sure there is at least 1 inch of water covering the jars.

9. Process for 10 minutes, adjusting for altitude. Remove the jars from the canner and cool.

Variation:

For a more Indian flavor, try Curried Mango Chutney. Omit the chili powder and substitute with 4 teaspoons of mild curry powder.

Peach Chutney

This chutney is a bit sweeter than traditional mango chutney, and it is wonderful on pork and poultry dishes. It also makes a delicious dipping sauce for shrimp and a unique condiment for wraps and sandwiches.

- 2 lemons, chopped finely with the peel and juice
- 4 cups apple cider vinegar
- 2 pounds packed dark brown sugar
- 2 cloves garlic, chopped
- 1 tablespoon mustard seed
- ¼ teaspoon cayenne pepper
- ¼ teaspoon chili powder
- ¼ teaspoon ground cinnamon
- 3 teaspoons pickling salt
- 2 serrano chili peppers, stemmed, seeded, and chopped
- 6 pounds firm, ripe peaches
- ¼ cup peeled and chopped fresh ginger
- ½ cup raisins
- 9 pint jars, lids, and bands

1. Place the chopped lemon and lemon juice in a large stockpot, and add the vinegar, sugar, garlic, mustard seed, cayenne pepper, chili powder, cinnamon, pickling salt, and chili peppers. Mix well.

2. Bring to a boil over medium-high heat, stirring frequently. Reduce the heat to medium and simmer for 30 minutes, stirring occasionally.

3. Bring a large pot of water to a boil, and use a metal colander to drop the whole peaches in. Boil for 30 seconds. Place the colander into a sink or large bowl of ice water to stop the cooking process. Slip the peels from the peaches, halve them, remove the stones, and then roughly chop the peaches.

4. Add the peaches, ginger, and raisins to the chutney mixture, stirring well, and continue to simmer for 40 to 45 minutes, stirring often, just until the peaches are tender and the syrup has thickened.

5. Fill the canner with enough water to cover the jars. Boil the water, reduce the heat to low, place the jars in the water, and simmer until ready to use.

6. Ladle the hot chutney into the canning jars, leaving ½ inch of headspace.

7. Remove air bubbles, wipe the rims, center the lids, and screw on the bands and adjust until they are fingertip tight.

8. Place the jars in the canner and bring to a boil. Make sure there is at least 1 inch of water covering the jars.

9. Process for 10 minutes, adjusting for altitude. Remove the jars from the canner and cool.

Firm, slightly tart peaches make the best peach chutney.

5

LOW-SODIUM AND LOW-SUGAR RECIPES

One of the many great things about doing your own canning is that you can control how much sugar and salt are included in your food. Most canned goods you buy at the store are loaded with sodium, high-fructose corn syrup, and other sweeteners. When you prepare your own canned foods, you have complete control.

Fruits and fruit spreads made without sugar or with very little sugar taste simply amazing, as long as you pick fruit that is at its peak flavor. Vegetables, sauces, and soups made with very little salt or even no salt at all can be absolutely delicious when the right herbs are used to flavor the dish.

Parents will appreciate the ability to can their own fresh fruits for the kids, without having to read the labels on overpriced packaged fruit snacks and without having to gasp at the sugar content.

In some of the included low-sugar recipes, it's advised that you use "low-sugar or no-sugar needed" pectin, as it will help your recipe thicken properly without needing quite as much sugar as other recipes. Any jam or jelly can be made without sugar, but some tend to have a texture more like gelatin. If that doesn't put you off, you can adapt your own jam and jelly recipes to low-sugar canning.

For people on low-sodium diets for medical reasons, home canning can be a huge help, allowing you to have a wide variety of foods in your pantry that meet your dietary guidelines and also taste wonderful.

Low-Sugar Apricot Freezer Jam

This delicious and simple apricot jam is a freezer recipe, so you'll need plastic canning containers made for the freezer. Although white grape juice is used in this recipe, you can also experiment with apple juice. This recipe also works very well for nectarines and peaches.

- 3 tablespoons low-sugar or no-sugar-needed pectin
- 1¾ cups unsweetened white grape juice
- 1 tablespoon lemon juice
- 3 cups (about 25 medium-sized) finely chopped fresh apricots

- 2–3 teaspoons granulated sugar, if desired
- 4 half-pint freezer canning containers with lids and bands

1. In a large saucepan, stir the pectin into the white grape juice and lemon juice until it's completely dissolved. Over medium-high heat, bring to a full boil, stirring frequently.

2. Boil the liquid hard for 1 minute, stirring constantly. Remove the pan from the heat.

3. Add the apricots into the hot pectin mixture and stir quickly for 1 minute. Stir in the sugar, if you're using it, until completely dissolved.

4. Ladle the hot jam into clean freezer containers, leaving ½ inch of headspace. Add the lids and bands, and let the jam sit in the refrigerator for up to 24 hours until set. This will keep in the refrigerator for 3 weeks or in the freezer for up to 1 year.

Pears Canned in Apple Juice

Home-canned fruits are a great way for parents to make sure kids always have a healthy and delicious snack on hand. This recipe is as simple as can be and far less expensive to make than to buy it in the stores.

- 6 pounds firm, ripe pears (Bartlett pears work very well)
- ½ –1 cup lemon juice
- 3 cups unsweetened apple juice
- 6 pint jars, lids, and bands

1. Fill the canner with enough water to cover the jars. Boil the water, reduce the heat to low, place the jars in the water, and simmer until ready to use.

2. Wash, peel, and core the pears, and cut them into halves lengthwise. Toss the pears in lemon juice (working in batches) to prevent browning, and set aside on a cutting board or clean surface.

3. In a small saucepan, bring the apple juice to a boil and then reduce the heat to medium to keep hot.

4. In a large stockpot, bring water to a boil, reduce the heat to medium, and cook the pears (working in single layer batches) just until heated through, about 1 to 2 minutes. Set each cooked batch of pears on paper towels to drain, and prepare to pack into the jars while still hot.

5. Pack the pears core side down, and pour in enough of the hot apple juice to cover while leaving ½ inch of headspace.

6. Remove air bubbles, wipe the rims, center the lids, and screw on the bands and adjust until they are fingertip tight.

7. Place the jars in the canner and bring to a boil. Make sure there is at least 1 inch of water covering the jars.

8. Process for 20 minutes, adjusting for altitude. Remove the jars from the canner and cool.

Low-Sugar Strawberry Jam

This incredibly simple recipe highlights the naturally sweet/tart flavor of ripe strawberries without adding a lot of sugar. This jam makes a wonderful gift, especially to families with young children or those on a low-sugar diet.

- 4 cups crushed fresh strawberries
- 1 cup unsweetened white grape juice

- 3 tablespoons low-sugar or no-sugar-needed pectin
- 6 half-pint jars, lids, and bands

1. Combine the strawberries and grape juice in a 6- to 8-quart saucepan. Gradually stir in the pectin until completely dissolved. Bring to a full rolling boil over high heat, stirring constantly.

2. Remove from the heat, and skim off any foam.

3. Ladle the hot jam into the canning jars, leaving ¼ inch of headspace.

4. Remove air bubbles, wipe the rims, center the lids, and screw on the bands and adjust until they are fingertip tight.

5. Place the jars in the canner and bring to a boil. Make sure there is at least 1 inch of water covering the jars.

6. Process for 20 minutes, adjusting for altitude. Remove the jars from the canner and cool.

Low-Sugar Peachy-Pineapple Spread

In this recipe, the sweetness of summer-ripened peaches blends beautifully with the pineapple. This fruit spread is a wonderful change of pace from typical berry jellies works beautifully as a glaze for ham or chicken. This recipe works with any combination of peaches, nectarines, apricots, and pineapple.

- 6 pounds firm, ripe peaches
- 2 cups chopped fresh pineapple
- ¼ cup lemon juice
- 6 half-pint jars, lids, and bands

1. Wash the peaches and drain well. Peel the fruit and remove the stones.

2. Place the peaches in a food processor, and pulse just a few times to mostly crush the fruit but not puree it. (You can also use a fork for this step.)

3. Place the peaches in a large saucepan over medium-low heat. Heat slowly while stirring to release the juice from the peaches, until the peaches are tender.

4. Transfer the peaches to a jelly bag or into a strainer lined with a few layers of cheesecloth. Place the jelly bag or strainer over a large bowl to catch the juices, which you can later use to can peaches or make peach jelly. Allow the peach juice to drip for about 15 minutes.

5. Fill the canner with enough water to cover the jars. Boil the water, reduce the heat to low, place the jars in the water, and simmer until ready to use.

6. Measure out 4 cups of the drained peach pulp for making the spread, and place into the empty saucepan. Add the pineapple and lemon juice, and mix well. Bring to a boil over medium-high heat, and boil gently for 10 to 15 minutes. Stir occasionally to prevent sticking.

7. Ladle the hot spread into the canning jars, leaving ½ inch of headspace.

8. Remove air bubbles, wipe the rims, center the lids, and screw on the bands and adjust until they are fingertip tight.

9. Place the jars in the canner and bring to a boil. Make sure there is at least 1 inch of water covering the jars.

10. Process for 20 minutes, adjusting for altitude. Remove the jars from the canner and cool.

Light Blueberry-Almond Fruit Spread

Blueberries and almond just naturally go together, and the flavors are highlighted in this low-sugar fruit spread. This tastes great spread on toast or muffins and also yummy stirred into plain or vanilla yogurt as a light snack. This is a freezer product, so you'll need plastic canning containers.

- 3 cups (about 2½ pints) blueberries
- 1 tablespoon lemon juice
- 1¾ cups unsweetened white grape juice
- 3 tablespoons low-sugar or no-sugar-needed pectin
- 1 teaspoon almond extract
- 4 half-pint freezer canning containers with lids and bands

1. Crush the blueberries in a large, shallow casserole dish using a potato masher or a large slotted spoon.

2. In a large saucepan, gradually stir the pectin into the grape juice and lemon juice until completely dissolved. Bring to a full rolling boil over medium-high heat, stirring frequently. Boil hard for 1 minute, stirring constantly. Remove from the heat.

3. Add the crushed blueberries into the hot pectin liquid, and stir vigorously for 1 minute. Stir in the almond extract and mix well.

4. Ladle the hot spread into the freezer containers, leaving ½ inch of headspace. Place in the refrigerator and let stand until set, up to 24 hours. You may keep one container in the refrigerator to eat immediately (for up to 3 weeks) and freeze the rest for up to 1 year.

Honey-Pear Jelly

Honey and pears are a wonderful combination. This jelly involves a little bit more work to make, but the end product is beautiful to look at, making it a wonderful holiday gift. Happily, it's as delicious as it is attractive.

- 12 medium just-ripened pears (Bartlett pears work well)
- 2 tablespoons lemon juice
- 3 cups water
- 3 tablespoons low-sugar or no-sugar-needed pectin
- ¾ cup honey
- 4 half-pint jars, lids, and bands

1. Core the pears and coarsely chop them, leaving the skins on. Place the pears in a large stockpot and add 3 cups of water. Simmer the pears for 10 minutes, covered, stirring occasionally.

2. Use a large spoon to place the pears in a damp jelly bag or in a strainer lined with several layers of damp cheesecloth. Let the juice drip for at least 2 hours or overnight. Don't squeeze the pears to make them drip faster; it will cause your jelly to be cloudy.

3. Fill the canner with enough water to cover the jars. Boil the water, reduce the heat to low, place the jars in the water, and simmer until ready to use.

4. Combine the pear juice and lemon juice in a large saucepan. Gradually stir in the pectin until completely dissolved. Bring the mixture to a full rolling boil over high heat, stirring constantly.

5. Add the honey and return the mixture to a full rolling boil. Boil hard for 1 minute, stirring constantly.

6. Remove the pan from the heat, and skim off any foam.

7. Ladle the hot jelly into the canning jars, leaving ¼ inch of headspace.

8. Remove air bubbles, wipe the rims, center the lids, and screw on the bands and adjust until they are fingertip tight.

9. Place the jars in the canner and bring to a boil. Make sure there is at least 1 inch of water covering the jars.

10. Process for 10 minutes, adjusting for altitude. Remove the jars from the canner and cool.

Honey-Pear Jelly

Low-Sodium Dill Pickles

Pickles don't need a lot of added salt to taste crisp and delicious. These pickles are so good that no one will even notice they're low in sodium.

- 4 pounds pickling cucumbers (3 to 4 inches long)
- 6 cups apple cider vinegar
- 2 tablespoons pickling salt
- 1½ teaspoons celery seed
- 1½ teaspoons mustard seed
- 2 large onions, thinly sliced
- 8 heads fresh dill
- 8 pint jars, lids, and bands

1. Trim the blossom ends of the cucumbers, cut into ¼-inch slices, and set aside.

2. Fill the canner with enough water to cover the jars. Boil the water, reduce the heat to low, place the jars in the water, and simmer until ready to use.

3. Combine the vinegar, pickling salt, and celery and mustard seeds in a large saucepan, and bring to a boil. Remove from the heat.

4. Place 2 slices of onion and ½ head of dill into the bottom of each pint jar.

5. Fill the jars with cucumber slices, leaving ½ inch of headspace. Top each jar with another slice of onion and half head of dill. Pour hot pickling solution over cucumbers, leaving ¼ inch of headspace.

6. Remove air bubbles, wipe the rims, center the lids, and screw on the bands and adjust until they are fingertip tight.

7. Place the jars in the canner and bring to a boil. Make sure there is at least 1 inch of water covering the jars.

8. Process for 20 minutes, adjusting for altitude. Remove the jars from the canner and cool.

Low-Sodium Sweet Pickle Slices

These simple sweet pickles are easy to make and easy to make disappear. Kids love them and they're a hit with adults as well.

- 4 pounds pickling cucumbers (3 to 5 inches long)
- 1⅔ cups plus 1 quart white vinegar, divided
- 3½ cups granulated sugar, divided

- 1 tablespoon whole allspice
- 2¼ teaspoons celery seed
- 1 tablespoon pickling salt
- 1 tablespoon mustard seed
- 4 pint jars, lids, and bands

1. Fill the canner with enough water to cover the jars. Boil the water, reduce the heat to low, place the jars in the water, and simmer until ready to use.

2. Trim the blossom ends of the cucumbers and cut into ¼-inch slices.

3. Combine 1⅔ cups of the vinegar, 3 cups of the sugar, allspice, and celery seed in a saucepan, and bring to a boil over medium-high heat, stirring until the sugar is completely dissolved. Reduce the heat to low and keep hot until ready to can.

4. In a large stockpot over medium-high heat, combine 1 quart of vinegar, the pickling salt, mustard seeds, and ½ cup of sugar, stirring until sugar is dissolved.

5. Add the cucumbers and cover. Bring to a boil, reduce the heat to medium, and simmer about 6 to 7 minutes, until the cucumbers change from bright to dull green.

6. Drain the cucumber slices and pack into the hot canning jars, leaving ½ inch of headspace. Add enough hot canning syrup to cover while leaving ¼ inch of headspace.

7. Remove air bubbles, wipe the rims, center the lids, and screw on the bands and adjust until they are fingertip tight.

8. Place the jars in the canner and bring to a boil. Make sure there is at least 1 inch of water covering the jars.

9. Process for 15 minutes, adjusting for altitude. Remove the jars from the canner and cool.

Low-Sodium Spicy Dilly Beans

Dilly beans are every bit as satisfying and addictive as salt and vinegar chips, but far better for you, especially if you're on a low-salt diet. These have very little salt but don't lack a bit of flavor.

- 2 pounds trimmed fresh green beans
- 2½ cups apple cider vinegar
- 2½ cups water
- 4 large garlic cloves
- 4 heads fresh dill
- 2 dried hot peppers, such as serrano, cut in quarters
- 4 pint jars, lids, and bands

1. Trim the green beans and wash very well. Drain on paper towels.

2. Fill the canner with enough water to cover the jars. Boil the water, reduce the heat to low, place the jars in the water, and simmer until ready to use.

3. Combine the vinegar and water in a large stockpot and bring to a boil.

4. Drop a garlic clove, one head of dill, and two pieces of dried pepper into each canning jar.

5. Pack the green beans into the jars, leaving ½ inch of headspace. Pour enough hot liquid over to cover while leaving ¼ inch of headspace.

6. Remove air bubbles, wipe the rims, center the lids, and screw on the bands and adjust until they are fingertip tight.

7. Place the jars in the canner and bring to a boil. Make sure there is at least 1 inch of water covering the jars.

8. Process for 10 minutes, adjusting for altitude. Remove the jars from the canner and cool.

Low-Sodium Zesty Pickled Peppers

These peppers are a great dish for cold buffets and antipasto platters, and as a topper for sandwiches, burgers, and hot dogs.

- 1 pound banana peppers, seeded and sliced into rings
- 2 large onions, sliced
- 4 cups apple cider vinegar
- 1¼ cups granulated sugar
- 2 tablespoons black peppercorns
- 1 teaspoon mustard seed
- 1 teaspoon celery seed
- 12 bay leaves
- Hot pepper sauce
- 6 half-pint jars, lids, and bands

1. Fill the canner with enough water to cover the jars. Boil the water, reduce the heat to low, place the jars in the water, and simmer until ready to use.

2. Combine the peppers, onions, vinegar, sugar, peppercorns, mustard seed, and celery seed in a large saucepan over high heat, and bring to a rolling boil. Boil hard for 1 minute and remove from the heat.

3. Drop 2 bay leaves into the bottom of each jar, and then fill with the pepper mixture, leaving ½ inch of headspace. Add three drops of hot pepper sauce to each jar. Pour enough liquid into each jar to cover while leaving ¼ inch of headspace.

4. Remove air bubbles, wipe the rims, center the lids, and screw on the bands and adjust until they are fingertip tight.

5. Place the jars in the canner and bring to a boil. Make sure there is at least 1 inch of water covering the jars.

6. Process for 10 minutes, adjusting for altitude. Remove the jars from the canner and cool.

SECTION 2

Pressure Canning

6

PRESSURE CANNING: SUPPLIES, INSTRUCTIONS, AND SAFETY

Using a pressure canner is the safe home-canning method for low-acid fruits and vegetables, as well as meats, poultry, and seafood, and prepared foods such as soups, stews, chilies, and pie fillings.

Preparing foods for pressure canning is very similar to preparing foods for water bath canning and uses many of the same tools and supplies. The difference is in the canning process itself. In water bath canning, you bring the water to a boil and process the jars for a specified amount of time. To safely process low-acid foods through pressure canning, you not only have to process the jars for a specified length of time but also under a specified pressure of pounds per square inch (psi).

Besides allowing you to preserve low-acid fruits and vegetables, pressure canning lets you take food preservation, meal readiness, and meal planning to a new level. Instead of just having your own home-canned fruits, jams, and pickles at hand, you can have entire meals, even a year's worth of meals, ready and waiting in your pantry.

If you're someone who gets a great deal of peace and security from knowing that you have a stockpile of food available at all times, pressure canning is going to be a huge help to you. Once you learn how easy it is to can almost anything, you'll be able to prepare, can, and store a month's worth of vegetables, soups, stews, and other meals in a single afternoon.

This can save you a tremendous amount of work, time, and money. Instead of spending a few hours preparing a pot of chicken soup, you can do just a little more prep, use another pot or two, and prepare and can enough chicken soup for a dozen, twenty, or even more meals. You save money by using far less energy, by being able to take advantage of especially cheap bulk purchases or sales, and by reducing the need for takeout or delivery meals on busy days.

Pressure canning intimidates many beginners, but you'll be pleasantly surprised at how easy it really is. After doing it once or twice, you'll wonder what took you so long to try it out.

What You'll Need for Pressure Canning

The tools and supplies for pressure canning are basically the same as those used for water bath canning. You'll use the same types of jars, lids, and rings, and you'll also need a jar lifter, plenty of clean towels, a butter knife or thin spatula, and one or two (or more) large stockpots. See below for an equipment checklist.

☑	Weighted- or dial-gauge pressure canner
☑	Rack for the bottom of the pressure canner
☑	Jar lifter (tongs used to lift hot jars from the canner)
☑	Jars (½ pint, pint, or quart sizes)
☑	Canning funnel
☑	New lids for each jar
☑	Bands for each jar
☑	Clean dish towels for wiping jar rims
☑	Clean dish towels for cooling the hot jars
☑	Butter knife or thin spatula for removing air bubbles
☑	Watch or kitchen timer
☑	Marker and labels

The main difference in your equipment is the canner. Several very good pressure canners are on the market, and you can also find very good pressure canners secondhand, with a few caveats that will be explained shortly. While a good pressure canner can cost you anywhere from $75 to $300, it will more than pay for itself in a single season. (Good pressure canners are built to last for decades.)

Pressure canners are made of stainless steel or heavy-cast aluminum, and they are made to be used on your stove top. (If you have a flat or solid surface stove top, check your manufacturer's literature to see what size pressure canners you can use for that model.) There are basically three types of pressure canners, and the difference is in how you see and regulate the pressure:

- A **dial-gauge pressure canner** has a dial that tells you what the current pressure is in your canner.

- With a **weighted-gauge canner**, a small piece over the steam vent jiggles or rocks when the correct pressure is being maintained.

- There are also **dual-gauge pressure canners**, which have both a dial and a weighted gauge, and the manufacturer's literature will tell you when and how to use each one.

These different pressure canners work equally well, and the one you purchase depends on your budget and personal preference.

Pressure canners also come in several different sizes, and those sizes, as with water bath canners, are indicated by how many quarts they can process at one time. Pressure canners come with at least one rack, although you can buy additional racks if needed. You usually use only one rack for canning quarts, but you may stack pint and half-pint jars in two or three layers by using additional racks.

USDA tested-recipes, which many of our recipes are based on, have been created and tested safe for canners large enough to accommodate a minimum of four quart-sized jars.

Should You Buy New or Secondhand?

Secondhand pressure canners can be found very easily through auctions, on sites like Freecycle and Craiglist, and also at garage sales or through friends and family members. While buying secondhand can represent a very nice savings, there are some cautions.

As a beginner, unless you are buying your pressure canner from an experienced canner you know and trust, you should stick with a new canner. Without experienced eyes, you may not be able to recognize flaws in the canner, such as damaged gaskets or seals, or broken or missing pieces from the gauge or canner. Also, pressure canners went through several redesigns for safety in the 1970s, and a canner made before then may not be as safe or as easy to use as a newer model (at least, not for an inexperienced home canner).

Once you have some experience under your belt and you want to purchase a larger or better quality pressure canner, a secondhand model can be a great option.

Basic Instructions for Pressure Canning

The following are the basic step-by-step instructions for canning with a pressure canner. These instructions are for both types of pressure canners. Always refer to the operating instructions for your canner, but these instructions can serve as an overall guide:

1. Read through the recipe you'll be using, and assemble all of the utensils, jars, lids, and bands you'll need. You also need plenty of clean hand or dish towels for wiping the rims of your jars and for cleaning up spills (and there will be spills). You want to gather all of your ingredients and have everything you'll be using (foods, spices, etc.) assembled and ready to go before you start.

2. Just as with water bath canning, you'll need to sterilize your jars and lids by putting them in hot water. Fill a large stockpot (or water bath canner) half full with water, and place the jars in the water before bringing it to a boil. Once it reaches a boil, turn the heat to medium, and allow them to simmer. Place the lids and bands in a small saucepan, and just simmer them.

3. To prepare your pressure canner, fill the canner with about 3 inches of water and set the canner on a burner. Turn the heat to medium-high, and allow the water to remain at a simmer while you're preparing the food to be canned. Once you have placed the filled jars in the canner, refer to your operator's manual for your specific type of canner's instructions.

4. Once you have prepared the food to be canned, use your jar lifter to remove a jar from the hot water. Dump out any water that may have entered the jar during heating. Use tongs to remove a lid from the saucepan. Fill the jar according to the recipe, being sure to leave the proper amount of headspace.

5. Use a butter knife or spatula to remove any air bubbles that may be in the jar after filling.

6. With a clean towel, carefully wipe the rims and threads around the top of the jar, removing any food or liquid that may have gotten on them. Always use a clean towel for this!

7. Center the lid onto the jar and then screw on the band, tightening just until it is fingertip tight.

8. Once you have a full rack's worth of jars (or fewer if you're making a small batch), place the jars into the canner using the jar rack. Make sure the canner has 2 to 3 inches of simmering water remaining inside.

9. Lock the canner lid into place, but leave the vent pipe open. Adjust the burner heat to medium-high. Allow the steam to escape through the vent pipe. Once you have a steady stream of steam escaping, vent for 10 minutes to make sure there is only steam in the canner, no air.

10. Close the vent using the weight or whichever method is used for your canner.

11. Per your operator's manual, gradually adjust the burner heat to reach and maintain the proper psi (pounds per square inch of pressure) called for in the recipe or instructions.

12. Follow your canner's manual and the recipe's instructions for processing the filled jars. The recipe and the instructions will tell you how long to process the jars and at what psi. Be sure to adjust for altitude, just as you do with water canning. See below for additional processing times needed for elevations over 1000 feet (305 meters). Also see the back of the book for the altitude of selected cities in the United States and Canada. To find the exact altitude of your location, use the search features on the EarthTools website (www.earthtools.org).

13. Once the required processing is completed, turn off the stove, remove the canner from the burner, and allow it to rest undisturbed until the gauge indicates that the pressure inside has returned to zero. Do not remove the weighted gauge!

14. Once you've reached zero pressure, wait an additional 2 minutes before following the manufacturer's instructions for venting and opening the canner. Always tilt the canner away from your face before removing the weight and unlocking the lid.

15. Remove the rack by its folding handles, and carefully set the jars on a clean towel. Use the jar lifter to remove individual jars from the rack.

16. Make sure the jars are in an area where they will be safe from jostling for the next 24 hours. (You may want to lay a dampened towel down first, with a dry towel on top. This will help keep the towel and jars more stable.)

17. Repeat these steps until all of the jars have been processed. Allow the jars to sit undisturbed for 24 hours before checking to make sure they have sealed properly.

18. To check the seals, unscrew the bands and press down on each lid with your finger. There should be no give or flex to the lid. The lid has a center similar to jar lids from store-bought foods; it will pop up slightly when the seal is broken. Your seals should all be flat. If any lids have popped, discard the food. You also should not be able to remove the lid by pulling lightly with your fingernails.

19. Once you are assured that all of the jars have sealed properly, label the jars with the date canned and the food within. You may want to put the date canned on the lid and the expected expiration date (usually 1 to 1½ years) on the label. This way, it will be easier for you to rotate jars (older to the front, newer to the back) as you do more canning.

20. Store the cooled and sealed jars in a cool, dark place where they're not likely to be disturbed.

Pressure Canning Altitude Chart

Altitude in Feet	Altitude in Meters	Weighted-Gauge Pounds / Dial-Gauge Pounds
0 – 1,000	0 – 305	10 / 11
1,001 – 2,000	306 – 609	15 / 11
2,001 – 4,000	610 – 1,219	15 / 12
4,001 – 6,000	1,220 – 1,829	15 / 13
6,001 – 8,000	1,830 – 2,438	15 / 14
8,001 – 10,000	2,439 – 3,048	15 / 15

Tips for Safety and Troubleshooting

Always follow your manufacturer's directions for using your specific model of pressure canner, but here are some tips for things to pay attention to as you're learning:

- Always make sure the correct amount of water is in the canner before you load your filled jars. Most foods require only about 2 to 3 inches of water in the canner, but foods with longer processing times (such as fish) may require more water at the start. Always read the directions to make sure.

- You must allow the pressure canner to vent fully for 10 minutes before you pressurize it. Once you have a steady flow of steam escaping from the vent, set a kitchen timer or alarm clock for 10 minutes. This ensures that there is no more air in the canner and that the proper psi can be reached once the canner is pressurized.

- Make sure you know the altitude where you live and that you always make adjustments for your altitude using the altitude chart in this book or in your manufacturer's instructions. The pressure and temperature inside the canner are lower at higher altitudes. If you don't adjust for the altitude, your food may not process properly and will not be safe to store and eat.

- Always check your gauges and seals for damage before you begin canning, and make sure the vent is clear of any food debris or hard water deposits.

- If at any time during processing your pressure goes below the required psi, bring the canner back to the correct pressure by raising the heat. Then start the timing of the process over again from the beginning.

- Never try to force the canner to cool faster than it will naturally. Trying to cool the canner with cold running water or by opening the vent before the canner is completely depressurized can cause a loss of liquid from the jars as well as seal failures. Forced cooling can warp your canner's lid, which means you'll have to replace the canner.

7

PRESSURE CANNING FRUITS AND VEGETABLES

Canning fruits and vegetables with a pressure canner is remarkably easy. Recipes are very simple, ingredients are few, and the flavor and texture of your produce usually will be much better than what's commonly found with commercially canned products.

There are two methods for canning most fruits and vegetables in a pressure canner: the raw-pack method and the hot-pack method:

- In the raw-pack method, raw fruit or vegetables are packed into the jars and then boiling water or other liquid is poured into the jars, leaving the prescribed amount of headspace.

- In the hot-pack method, the fruit or vegetables are boiled, with or without liquid, and then packed into the jars.

In many recipes you can use either method of packing. When one method is superior to another, the method to use is specified in the recipes. Very often, the choice will be yours, especially as you gain experience and start canning other foods and using other recipes.

One of the many advantages of pressure canning your own fruits and vegetables is that you can control the amount of sugar, salt, or both. Fruits can be canned in water with or without sugar, and in fruit juice without any added sugar. Vegetables can be canned with or without salt, and you may choose to use less salt than is called for in the recipe.

Always choose the freshest produce possible, and check carefully for spoilage, bruising, cuts, holes, or mold before preparing the produce for canning.

Although tomatoes may be canned using either water bath canning or pressure canning, this section includes the tomato recipes. Tomatoes vary from type to type regarding the amount of acid they contain. Since this book is intended for beginning canners, including tomatoes in this section saves you the process of determining the acid level of your tomatoes and deciding which method to use. The taste and texture of the tomatoes really doesn't vary between methods, so it's best to start with pressure canning tomatoes. As a bonus, pressure canning results in higher quality and more nutritious canned tomatoes. (If your pressure canner does not go above 15 psi, choose a longer process time at a lower pressure from a reliable chart.)

You should always check altitude charts and your manufacturer's directions for processing times and psi. They will differ between dial-gauge and weighted-gauge canners, and also depend on whether you are canning pint jars or quart jars. For simplicity's sake, the recipes in this book specify the processing times and psi for using dial-gauge pressure canners.

Carrots

Always use the freshest carrots you can find for canning. Dried-out carrots are usually fibrous and have lost their natural sweetness. Rubbery carrots will have a soft and mealy texture after canning. Smaller carrots that are less than 1 to 1½ inches in diameter are best. These carrots can even be canned whole, which is very attractive and saves you time. Larger carrots should be sliced or cut into chunks. You may want to alternate between slices and chunks to accommodate your favorite recipes.

- 11 pounds fresh carrots
- 4½ teaspoons pickling salt
- 9 pint jars, lids, and bands

1. Prepare your jars by bringing them to a boil in a large pot. Once they reach a boil, reduce the heat to low, and allow them to simmer until you're ready to use them. Prepare the lids and bands by simmering (not boiling) them over low heat in a small saucepan.

2. Prepare your canner by filling it with 2 to 3 inches of water and bringing it to a boil over high heat.

3. Fill another large pot with water and bring to a boil. Continue boiling until ready to use.

4. Wash and trim the ends from the carrots. Peel and then rewash. Slice or cut the carrots as desired.

5. For raw pack: Pack the carrots into the jars, leaving 1 inch of headspace. Once all of the jars are packed, pour in enough boiling water to cover, leaving ½ inch of headspace.

6. For hot pack: Place the carrots into the boiling water, reduce the heat to medium, and simmer for 5 minutes. Use a slotted spoon to pack the carrots into the jars, leaving 1 inch of headspace. Add enough of the cooking water to cover, leaving 1 inch of headspace.

7. Add ½ teaspoon of pickling salt to each jar.

8. Remove air bubbles, wipe rims, center the lids, and screw on the bands and adjust until they are fingertip tight. Place the jars in the canner.

9. Cover, vent, and pressurize the canner according to its manufacturer's directions.

10. For both raw- and hot-pack methods, process the jars for 25 minutes at 11 pounds of pressure, adjusting for altitude. Remove the canner from the burner, and follow the directions for handling noted in "Basic Instructions for Pressure Canning."

Green Beans

You can use either bush beans or pole beans for canning, but you'll want young beans of either type. Older, larger beans may be too fibrous and lack sweetness and flavor. Beans can be canned whole, cut into pieces, or cut French style.

- 9 pounds fresh green beans
- 4½ teaspoons pickling salt

- 9 pint jars, lids, and bands

1. Prepare your jars by bringing them to a boil in a large pot. Once they reach a boil, reduce the heat to low, and allow them to simmer until you're ready to use them. Prepare the lids and bands by simmering (not boiling) them over low heat in a small saucepan.

2. Prepare your canner by filling it with 2 to 3 inches of water and bringing it to a boil over high heat.

3. Fill another large pot with water and bring to a boil. Continue boiling until ready to use.

4. Wash the green beans and trim off the ends. If there are strings, remove them with your fingertips or a small paring knife.

5. For raw pack: Pack the beans into the jars, leaving 1 inch of headspace. Once all of the jars are packed, pour in enough boiling water to cover, leaving 1 inch of headspace.

6. For hot pack: Place the beans into the boiling water, reduce the heat to medium and simmer for 5 minutes. Use a slotted spoon to pack into jars, leaving 1 inch of headspace. Add enough of the cooking water to cover, leaving 1 inch of headspace.

7. Add ½ teaspoon of pickling salt to each jar.

8. Remove air bubbles, wipe rims, center the lids, and screw on the bands and adjust until they are fingertip tight. Place the jars in the canner.

9. Cover, vent, and pressurize the canner according to its manufacturer's directions.

10. For both raw- and hot-pack methods, process the jars for 20 minutes at 11 pounds of pressure, adjusting for altitude. Remove the canner from the burner, and follow the directions for handling noted in "Basic Instructions for Pressure Canning."

Variation:

For Italian Green Beans, add 1 teaspoon of fresh oregano and 1 whole garlic clove to each jar before adding the liquid.

Tomatoes

Choose very ripe tomatoes for canning. They'll be the sweetest and most flavorful. To remove the peels, plunge the tomatoes into boiling water for 30 seconds to 1 minute, just until the skins begin to split. Immediately plunge them into ice-cold water, off the skins, and then core them (and cut them if needed).

- 13 pounds peeled tomatoes
- 4½ teaspoons pickling salt
- 4½ teaspoons granulated sugar
- 9 tablespoons lemon juice
- 9 pint jars, lids, and bands

1. Prepare your jars by bringing them to a boil in a large pot. Once they reach a boil, reduce the heat to low, and allow them to simmer until you're ready to use them. Prepare the lids and bands by simmering (not boiling) them over low heat in a small saucepan.

2. Prepare your canner by filling it with 2 to 3 inches of water and bringing it to a boil over high heat.

3. Fill another large pot with water and bring to a boil. Continue boiling until ready to use.

4. If you prefer, you can halve the tomatoes. Otherwise, leave them whole.

5. For raw pack: Pack the tomatoes into the jars, leaving ½ inch of headspace. Once all of the jars are packed, pour in enough boiling water to cover, leaving ½ inch of headspace.

6. For hot pack: Place the tomatoes into the boiling water, reduce the heat to medium, and simmer for 5 minutes. Pack into jars, leaving 1 inch of headspace. Add enough of the cooking water to cover, leaving 1 inch of headspace.

7. Add ½ teaspoon of salt, ½ teaspoon of sugar, and 1 tablespoon of lemon juice to each pint.

8. Remove air bubbles, wipe rims, center the lids, and screw on the bands and adjust until they are fingertip tight. Place the jars in the canner.

9. Cover, vent, and pressurize the canner according to its manufacturer's directions.

10. For both raw- and hot-pack methods, process the tomatoes for 15 minutes at 6 pounds of pressure, adjusting for altitude. Remove the canner from the burner, and follow the directions for handling noted in "Basic Instructions for Pressure Canning."

Variations:

You can also prepare Tomatoes Canned in Tomato Juice. Bring 2 quarts of tomato juice to a boil and for raw pack, pour over the raw, packed tomatoes, leaving ½ inch of headspace. For hot pack, simmer the tomatoes in the tomato juice for 5 minutes; add the salt, sugar, and lemon juice to each jar; and then pack the tomatoes into jars, leaving ½ inch of headspace.

For Crushed Tomatoes, you'll need 14 pounds of tomatoes. After peeling and coring, cut the tomatoes into quarters. Crush one-fourth of the tomatoes in a large stockpot (a potato masher works well), and bring to a boil over medium-high heat. Once boiling, reduce the heat to medium and add the remaining tomatoes. Cook for 5 minutes; add the salt, sugar, and lemon juice to each jar; and then pack according to the hot-pack directions given earlier.

Whole-Kernel Corn

Home-canned corn is far more flavorful than the commercially canned versions. For the absolute best flavor, choose corn that has just been picked, either from your garden, the garden of a friend, or from a local farm or farmers' market. The sugars in corn begin to convert to starch as soon as it's picked.

• 20 pounds (about 40 ears) fresh, raw corn	• 4½ teaspoons pickling salt
	• 9 pint jars, lids, and bands

1. Prepare your jars by bringing them to a boil in a large pot. Once they reach a boil, reduce the heat to low, and allow them to simmer until you're ready to use them. Prepare the lids and bands by simmering (not boiling) them over low heat in a small saucepan.

2. Prepare your canner by filling it with 2 to 3 inches of water and bringing it to a boil over high heat.

3. Fill another large pot with water and bring to a boil. Add the whole ears of corn (in batches if needed) and blanch for 3 minutes. Remove from the water and place in a bowl of cold water.

4. Using a corn cobber or sharp paring knife, stand an ear of corn up with the wide end at the bottom, holding it firmly at the narrow end. Cut the rows of kernels off about three-quarters of the way to the cob. Do not scrape the cob. As you go, swipe the cut kernels into a large container. Repeat with each cob.

5. For raw pack: Once all of the corn has been cut, bring another large pot of water to a boil and allow it to continue boiling until you're ready to use it. Loosely pack the corn into the jars, leaving 1 inch of headspace. Add ½ teaspoon of pickling salt to each jar. Pour in enough boiling water to cover, leaving 1 inch of headspace.

6. For hot pack: Use a quart jar or measuring pitcher to measure the corn kernels. Put all of the corn into a large pot, and for every quart of corn, add 1 cup of cold water. Bring to a boil and cook for 5 minutes. Pack the corn loosely into the jars, leaving 1 inch of headspace. Add ½ teaspoon of pickling salt to each jar. Pour in enough of the cooking water to cover, leaving 1 inch of headspace.

7.   Remove air bubbles, wipe rims, center the lids, and screw on the bands and adjust until they are fingertip tight. Place the jars in the canner.

8.   Cover, vent, and pressurize the canner according to its manufacturer's directions.

9.   For both raw- and hot-pack methods, process the corn for 55 minutes at 11 pounds of pressure, adjusting for altitude. Remove the canner from the burner, and follow the directions for handling noted in "Basic Instructions for Pressure Canning."

Variation:

For Cream-Style Corn, use a large bowl and stand the blanched ears of corn up in the bowl. Cut the kernels from the cobs, cutting all the way to the cob. Once all of the corn has been cut, scrape the "milk" from the cob with the back of a knife, catching it in the bowl.

For each quart of corn and scrapings you place in a large pot, add 2 cups of water. Bring to a boil and remove from heat. Add ½ teaspoon of salt to each jar, and pack the corn loosely into the jars, adding enough of the cooking liquid to cover, leaving 1 inch of headspace. Process for 85 minutes at 11 pounds of pressure, adjusting for altitude.

Asparagus Spears

Asparagus is so delicious and has such a short harvest window that it's considered the prize of most gardens. Canning is a great way to have asparagus all year round, without having to pay the high prices of out-of-season asparagus. You can use either large or small spears, but make sure the buds are tightly closed—this means they're fresh and young.

- 16 pounds trimmed and peeled asparagus spears
- 4½ teaspoons pickling salt
- 9 pint jars, lids, and bands

1. Prepare your jars by bringing them to a boil in a large pot. Once they reach a boil, reduce the heat to low, and allow them to simmer until you're ready to use them. Prepare the lids and bands by simmering (not boiling) them over low heat in a small saucepan.

2. Prepare your canner by filling it with 2 to 3 inches of water and bringing it to a boil over high heat.

3. Fill another large pot with water and bring to a boil. Continue boiling until ready to use.

4. To prepare the asparagus, trim off the woody stems just at the point at which they can no longer be bent. If the asparagus spears are very large, you may want to peel them gently with a vegetable peeler to remove tougher skin.

5. For raw pack: Add ½ teaspoon of pickling salt to each jar and pack the spears, leaving 1 inch of headspace. Once all of the jars are packed, pour in enough boiling water to cover, leaving 1 inch of headspace.

6. For hot pack: Place the spears into the boiling water, reduce the heat to medium, and simmer for 2 minutes. Add ½ teaspoon of pickling salt to each jar and pack the spears, leaving 1 inch of headspace. Add enough of the cooking water to cover, leaving 1 inch of headspace.

7. Remove air bubbles, wipe rims, center the lids, and screw on the bands and adjust until they are fingertip tight. Place the jars in the canner.

8. Cover, vent, and pressurize the canner according to its manufacturer's directions.

9. For both raw- and hot-pack methods, process the asparagus for 30 minutes at 11 pounds of pressure, adjusting for altitude. Remove the canner from the burner, and follow the directions for handling noted in "Basic Instructions for Pressure Canning."

Potatoes

While you can use larger potatoes for canning, the texture will be much better if you use new potatoes. They have a firmer texture with less starch than regular potatoes, so the flavor and texture will be far superior. If you use large potatoes, it's best to halve or dice them. New potatoes can be left whole if quite small or halved if you prefer. Potatoes are canned only with the hot-pack method. Note: To prevent the potatoes from discoloring, they need to be treated with a citric-acid solution, using either citric acid powder or 500 mg vitamin C tablets. For citric acid, 1 level teaspoon (about 3 grams) is needed per gallon of water. If using vitamin C tablets, use 6 tablets per gallon of water, crushing them before adding them to the water.

- 13 pounds washed and peeled potatoes
- Citric acid solution (see earlier note)
- 4½ teaspoons pickling salt
- 9 pint jars, lids, and bands

1. Prepare your jars by bringing them to a boil in a large pot. Once they reach a boil, reduce the heat to low, and allow them to simmer until you're ready to use them. Prepare the lids and bands by simmering (not boiling) them over low heat in a small saucepan.

2. Prepare your canner by filling it with 2 to 3 inches of water and bringing it to a boil over high heat.

3. Place the potatoes in a large bowl or enamel pot containing the citric acid solution. Let stand for 5 minutes, and then drain.

4. Bring a large pot of water to a boil and add the potatoes. Cook diced or cut potatoes for 2 minutes and whole new potatoes for 10 minutes. Drain.

5. Meanwhile, bring a clean pot of water to a boil.

6. Add ½ teaspoon of pickling salt to each jar and pack in the potatoes, leaving 1 inch of headspace. Pour in enough of the boiling water to cover, leaving 1 inch of headspace.

7. Remove air bubbles, wipe rims, center the lids, and screw on the bands and adjust until they are fingertip tight. Place the jars in the canner.

8. Cover, vent, and pressurize the canner according to its manufacturer's directions.

9. Process for 35 minutes at 11 pounds of pressure, adjusting for altitude. Remove the canner from the burner, and follow the directions for handling noted in "Basic Instructions for Pressure Canning."

Peaches

Home-canned peaches are so superior to commercially canned ones that they're one of the most worthwhile fruits to can at home. Choose firm, but ripe, peaches free of bruising or blemishes. You can use either white or yellow peaches. Because of their delicate texture, most canners prefer the raw-pack method for peaches. This recipe also works great for nectarines.

- 11 pounds firm, ripe peaches
- 6½ cups water

- ¾ cup granulated sugar
- 9 pint jars, lids, and bands

---

1. Prepare your jars by bringing them to a boil in a large pot. Once they reach a boil, reduce the heat to low, and allow them to simmer until you're ready to use them. Prepare the lids and bands by simmering (not boiling) them over low heat in a small saucepan.

2. Prepare your canner by filling it with 2 to 3 inches of water and bringing it to a boil over high heat.

3. Fill another large pot with water and bring to a boil. Add the peaches to the boiling water and boil for 30 seconds. Plunge into a large bowl of icy cold water to loosen the skins. Slip off the skins, and then cut the peaches in half and remove the pits.

4. Pack the peaches into the jars, leaving ½ inch of headspace.

5. In a heavy saucepan, combine the water and sugar and bring to a boil, stirring to dissolve the sugar. Pour enough syrup over the peaches to cover, leaving ½ inch of headspace.

6. Remove air bubbles, wipe rims, center the lids, and screw on the bands and adjust until they are fingertip tight. Place the jars in the canner.

7. Cover, vent, and pressurize the canner according to its manufacturer's directions.

8. Process for 10 minutes at 6 pounds of pressure, adjusting for altitude. Remove the canner from the burner, and follow the directions for handling noted in "Basic Instructions for Pressure Canning."

Variations:

To make Spiced Peaches, add half a cinnamon stick and 3 whole cloves to each jar before adding the hot syrup.

To make Peaches Canned in Juice, substitute 7 cups of apple or white grape juice for the water and sugar and bring to a boil. Pour over the jarred peaches and continue the preceding instructions.

Apples

Apples can vary nicely and are a great way to take advantage of an abundant harvest—either yours or someone else's! Choose firm apples that are both tart and sweet. Granny Smith, Crispin, and Yellow Delicious are good examples.

- 12½ pounds apples
- Citric acid solution (see note for **Potatoes** recipe)
- 2½ pints apple juice (you may substitute 2½ cups of the light syrup found in the **Peaches** recipe)
- 9 pint jars, lids, and bands

1. Prepare your jars by bringing them to a boil in a large pot. Once they reach a boil, reduce the heat to low, and allow them to simmer until you're ready to use them. Prepare the lids and bands by simmering (not boiling) them over low heat in a small saucepan.

2. Prepare your canner by filling it with 2 to 3 inches of water and bringing it to a boil over high heat.

3. To prepare the apples for canning, core, peel, and slice them into 6 to 8 pieces per apple. Place in the citric acid solution while you prepare the juice or syrup.

4. In a small saucepan, bring the juice or syrup to a boil.

5. Drain the apples and place in a large stockpot over medium heat. Cover with the juice or syrup. Cook for 2 minutes, stirring frequently to prevent sticking.

6. Pack the apples into the jars, leaving ½ inch of headspace. Cover with hot juice or syrup, leaving ½ inch of headspace.

7. Remove air bubbles, wipe rims, center the lids, and screw on the bands and adjust until they are fingertip tight. Place the jars in the canner.

8. Cover, vent, and pressurize the canner according to its manufacturer's directions.

9. Process for 8 minutes at 6 pounds of pressure, adjusting for altitude. Remove the canner from the burner, and follow the directions for handling noted in "Basic Instructions for Pressure Canning."

Applesauce

Applesauce

Making your own applesauce is fun and very rewarding . . . and not as much work as you might expect. Choose firm, slightly tart apples such as Granny Smith, Crispin, and Yellow Delicious, and feel free to mix them up a bit to vary the flavor.

Prepare the apples by peeling, coring, and slicing them, then soak them in a citric acid solution, either citric acid powder or 500 mg vitamin C tablets. For citric acid, 1 level teaspoon (about 3 grams) is needed per gallon of water. If using vitamin C tablets, use 6 tablets per gallon of water, crushing them before adding them to the water.

- 13½ pounds fresh apples
- Citric acid solution (see note for **Potatoes** recipe)
- ½ cup cold water
- ¼ cup granulated sugar
- ½ teaspoon salt
- 9 pint jars, lids, and bands

---

1. Prepare your jars by bringing them to a boil in a large pot. Once they reach a boil, reduce the heat to low, and allow them to simmer until you're ready to use them. Prepare the lids and bands by simmering (not boiling) them over low heat in a small saucepan.

2. Prepare your canner by filling it with 2 to 3 inches of water and bringing it to a boil over high heat.

3. To prepare the apples for canning, core, peel, and slice them into 6 to 8 pieces per apple. Place in the citric acid solution.

4. Drain the apple slices and place in a large stockpot over medium heat. Add the water, sugar, and salt.

5. Cook for 10 to 20 minutes, depending on the variety, just until tender. Stir frequently to prevent sticking.

6. Press through a sieve or food mill, or leave them alone if you like a chunkier applesauce.

7. Pack into the jars, leaving ½ inch of headspace.

8. Remove air bubbles, wipe rims, center the lids, and screw on the bands and adjust until they are fingertip tight. Place the jars in the canner.

9. Cover, vent, and pressurize the canner according to its manufacturer's directions.

10. Process for 8 minutes at 6 pounds of pressure, adjusting for altitude. Remove the canner from the burner, and follow the directions for handling noted in "Basic Instructions for Pressure Canning."

Variation:

For Spiced Applesauce, add 1 tablespoon of cinnamon, 1 teaspoon of nutmeg, and 1 teaspoon of allspice to the apples before cooking.

Pears

Canned pears are simple to make and wonderful to eat. The attractiveness of the finished product makes pears a great gift to give as a holiday or hostess gift. Bartlett and Bosc pears are good varieties for canning.

- 11 pounds firm ripe pears
- Citric acid solution (see note for **Potatoes** recipe)

- 4 cups white grape or apple juice

1. Prepare your jars by bringing them to a boil in a large pot. Once they reach a boil, reduce the heat to low, and allow them to simmer until you're ready to use them. Prepare the lids and bands by simmering (not boiling) them over low heat in a small saucepan.

2. Prepare your canner by filling it with 2 to 3 inches of water and bringing it to a boil over high heat.

3. Prepare the pears by peeling, coring, and slicing them in half lengthwise; then soak them in a citric acid solution.

4. In a separate large pot, combine the pears with the grape or apple juice (enough to cover, refrigerate any extra) and bring to a boil. Cook, stirring frequently, for 5 minutes.

5. Pack the pears into the jars and pour in enough of the hot juice to cover, leaving ½ inch headspace.

6. Remove air bubbles, wipe rims, center the lids, and screw on the bands and adjust until they are fingertip tight. Place the jars in the canner.

7. Cover, vent, and pressurize the canner according to its manufacturer's directions.

8. Process for 10 minutes at 6 pounds of pressure, adjusting for altitude. Remove the canner from the burner, and follow the directions for handling noted in "Basic Instructions for Pressure Canning."

Variation:

To make Vanilla Pears, add 1 teaspoon of nutmeg to the pears before cooking. When they have finished cooking, stir in 2 teaspoons of vanilla extract, mixing well. Vanilla Pears make a great topper for ice cream!

8

PRESSURE CANNING MEAT, POULTRY, AND SEAFOOD

Most people think of freezing when laying up a stock of meat and seafood. However, canning is a great way to preserve fresh meats and leave them tender and juicy. It's also a much simpler process than you might imagine.

All meats, seafood, and poultry products must be canned in a pressure canner, although you can use either a raw-pack or hot-pack method for most of them. When either is an option, instructions for both are included.

Chapter 9 features recipes for making prepared foods and meal starters, many of which contain meat, poultry, or seafood. This chapter, however, sticks with plain meats that you can use later for any recipe you like. For this reason, what follows are instructions, rather than recipes.

Remember to consult your altitude guide and the manufacturer's instructions for your specific pressure canner to adjust processing time and the proper pressure for the food you are canning. The times and pressures listed here are for dial-gauge pressure canners.

Processing Chicken, Turkey, and Game Birds

You can process poultry either with or without the bones, as long as you consult the recipe for proper processing time and pressure, as they differ between boneless and bone-in meats.

Because you can easily pack meats into either pint or quart jars and may choose to debone the meat or leave it on the bone, quantities aren't specified here. Simply choose which size jar will best suit your typical recipe needs, and have plenty of jars that size prepared for use. After processing poultry a few times, you'll have a good idea how many pounds of a given cut (both boneless and bone-in) you can pack into a quart or pint jar.

Always choose the freshest possible chicken and poultry for canning and rinse it very well. Then pat dry with paper towels.

1. Prepare your jars by bringing them to a boil in a large pot. Once they reach a boil, reduce the heat to low, and allow them to simmer until you're ready to use them. Prepare the lids and bands by simmering (not boiling) them over low heat in a small saucepan.

2. Prepare your canner by filling it with 2 to 3 inches of water and bringing it to a boil over high heat.

3. For the raw-pack method: Bring a large stockpot of water to a boil. Pack the raw meat into the jars and add enough boiling water to cover, leaving 1¼ inches of headspace. Add ½ teaspoon of salt to each jar.

4. For the hot-pack method: Boil, steam, or bake the poultry until it is about two-thirds done. (If baking or steaming, bring a large pot of water to a boil and allow it to boil until ready for use.)

5. Pack the poultry loosely into the jars and add enough boiling water to cover, leaving 1¼ inches of headspace.

6. Add ½ teaspoon of salt per pint and 1 teaspoon of salt per quart.

7. Remove air bubbles, wipe rims, center the lids, and screw on the bands and adjust until they are fingertip tight. Place the jars in the canner.

8. Cover, vent, and pressurize the canner according to its manufacturer's directions.

9. Processing times are the same for hot-pack and raw-pack methods, but vary between boneless and bone-in meats.

10. For boneless poultry: Pints should be processed for 65 minutes at 11 pounds of pressure and quarts should be processed for 75 minutes at 11 pounds of pressure, adjusting for altitude.

11. For bone-in poultry: Pints should be processed for 75 minutes at 11 pounds of pressure and quarts for 90 minutes at 11 pounds of pressure, adjusting for altitude.

12. Remove the canner from the burner, and follow the directions for handling noted in "Basic Instructions for Pressure Canning."

Processing Ground Meats

Ground meats are some of the easiest meats to can at home, and their flexibility makes them among the most popular, as well. The leaner mixes of ground beef, such as 80 to 85 percent lean, are better for canning, as higher fat content can alter the flavor. Be sure to drain as much fat as you can from all meats after cooking and before packing them into jars. Keep the following in mind:

- Even after draining your meat, the cooled canned product will develop an opaque, cloudy surface. This is just the cooled fat and is perfectly normal and acceptable. It will melt when you reheat the meat.

- Pack ground meats loosely; do not pack them down, as the liquid needs to be able to get to all of the meat.

- Ground meats are canned only with the hot-pack method.

1. Prepare your jars by bringing them to a boil in a large pot. Once they reach a boil, reduce the heat to low, and allow them to simmer until you're ready to use them. Prepare the lids and bands by simmering (not boiling) them over low heat in a small saucepan.

2. Prepare your canner by filling it with 2 to 3 inches of water and bringing it to a boil over high heat.

3. In another large pot, bring a full pot of water to a boil, and let it boil until ready to use. If you prefer to pack the meat in broth, use enough broth to cover the meat in the jars, adding (and bringing to a boil) more broth as needed.

4. Shape the ground meat into patties or balls if desired, or you may brown the meat without shaping it.

5. Sauté the meat until lightly browned, pack into the jars, and add enough boiling water or broth to cover, leaving 1¼ inches of headspace.

6. Add ½ teaspoon of salt per pint and 1 teaspoon of salt per quart.

7. Remove air bubbles, wipe rims, center the lids, and screw on the bands and adjust until they are fingertip tight. Place the jars in the canner.

8. Cover, vent, and pressurize the canner according to its manufacturer's directions.

9. Process pints for 75 minutes at 11 pounds of pressure and quarts for 90 minutes at 11 pounds of pressure, adjusting for altitude. Remove the canner from the burner, and follow the directions for handling noted in "Basic Instructions for Pressure Canning."

Processing Stew Meats

Chunks, cubes, or strips of meat can be processed using either the raw-pack or hot-pack methods. They work equally well, so the one you choose should be based on your cooking habits, what recipes you will be likely to use the meats for, and which method makes you feel most comfortable.

1. Prepare your jars by bringing them to a boil in a large pot. Once they reach a boil, reduce the heat to low, and allow them to simmer until you're ready to use them. Prepare the lids and bands by simmering (not boiling) them over low heat in a small saucepan.

2. Prepare your canner by filling it with 2 to 3 inches of water and bringing it to a boil over high heat.

3. Fill another large pot with water and bring to a boil. Continue boiling until ready to use. If you prefer, you may use stock or broth instead. If you choose to stew your meat, use the cooking liquid for extra flavor when packing. If you roast or brown the meat, you can add the drippings to the jar if you like.

4. For hot pack: Precook the meat to rare by roasting, stewing, or browning it. Pack loosely into the jars and add enough of the hot liquid to cover, leaving 1 inch of headspace.

5. For raw pack: Fill the jars loosely with raw meat pieces, leaving 1 inch of headspace. Do not add liquid.

6. Add ½ teaspoon of salt per pint and 1 teaspoon of salt per quart.

7. Remove air bubbles, wipe rims, center the lids, and screw on the bands and adjust until they are fingertip tight. Place the jars in the canner.

8. Cover, vent, and pressurize the canner according to its manufacturer's directions.

9. Processing is the same for both methods of packing. Process pints for 75 minutes at 11 pounds of pressure and quarts for 90 minutes at 11 pounds of pressure, adjusting for altitude. Remove the canner from the burner, and follow the directions for handling noted in "Basic Instructions for Pressure Canning."

Processing Fish

Canning fish may seem a little daunting at first, but it's a great way to preserve a bounty of fish that you've caught yourself or have been given.

Bluefish, mackerel, salmon, steelhead, trout, and other fatty fish are all good options for canning, with the exception of tuna. If you do decide to can salmon, be aware that magnesium ammonium phosphate sometimes forms crystals in canned salmon. This is harmless and will usually dissolve when you heat the fish.

For best practices, always gut the fish within 2 hours of catching it and keep it on ice until you're ready to use it.

The following instructions are for canning fish in pint jars:

1. Prepare your jars by bringing them to a boil in a large pot. Once they reach a boil, reduce the heat to low, and allow them to simmer until you're ready to use them. Prepare the lids and bands by simmering (not boiling) them over low heat in a small saucepan.

2. Prepare your canner by filling it with 2 to 3 inches of water and bringing it to a boil over high heat.

3. Remove the head, tail, fins, and scales. These are wonderful for making fish stock but should not be canned. Wash the fish, pat it dry, and butcher it into pieces no more than 3½ inches in length.

4. Loosely pack the fish into the jars, skin side against the glass, leaving 1 inch of headspace. Do not add any liquid.

5. Add 1 teaspoon of salt per jar, if desired.

6. Wipe rims, center the lids, and screw on the bands and adjust until they are fingertip tight. Place the jars in the canner.

7. Cover, vent, and pressurize the canner according to its manufacturer's directions.

8. Process for 100 minutes at 11 pounds of pressure, adjusted for altitude. Remove the canner from the burner, and follow the directions for handling noted in "Basic Instructions for Pressure Canning."

9

PRESSURE CANNING PREPARED FOODS AND MEAL STARTERS

Once you get the hang of canning simple items, you'll probably be eager to explore the many ways you can use canning to make meal planning easier and more fun, as well as to make feeding your family less time consuming and expensive. Canning your own prepared foods and meal starters does all that and more.

Just about any meal your family enjoys, including your favorite soups, stews, and chilies, may be preserved through home canning. While canning your own meats, vegetables, and fruits gives you all the components of a great meal, canning whole entrées in a single jar means that dinner or a great dessert is always right at your fingertips.

You can also start making large batches of meal starters. These are things that you use frequently for preparing many of your favorite meals, including your favorite pasta sauces, taco and fajita fillings, pie fillings, or seasoned ground meats that are the foundation of many casseroles.

The beauty of canning prepared foods and meal starters is that they can save you an enormous amount of time and money. It doesn't take much more time to prepare and can twelve jars of spaghetti sauce than it does to cook one pot of sauce for a single meal.

Canning your own prepared foods also means that you can take advantage of great sales and bulk purchases when prices are at their lowest. If ground beef is on sale for $1.99 per pound, you make a large purchase and can the meat alone or with batches of spaghetti sauce, taco meat, or minestrone.

Having an assortment of meals that can be reheated in minutes also means you don't have to spend a fortune ordering pizza, going to the drive-through, or running to a restaurant on nights when you're especially busy or tired, or when you've forgotten to thaw something for dinner.

No matter how you look at it, canning your own meals and meal starters saves you time and money and is incredibly satisfying.

Tips to Make Meal Planning Easier with Canning

- The best way to begin your meal-canning adventure is to think about what your family typically eats in a given month. What are your favorite meals? What are your go-to entrées when you haven't planned ahead? Do you usually like to whip up some sloppy joes, a quick chicken soup, or some spaghetti? Which meals show up time and time again from one week to another? Do you eat a lot of chicken casseroles or meat stews?

- Those are the meals that you need to start with when canning prepared foods and meal starters. They should become your foundational meals. You already know the recipes, and everyone in the family enjoys eating them. You know what you need to have on hand, and you can probably easily figure out how to double, triple, and quadruple the measurements.

- With meal starters, you can cut down on preparation time for many of your favorite meals. When you have homemade spaghetti sauce, seasoned taco meat, or chicken with gravy on your pantry shelf, you're already halfway to dinner before you even pull into the driveway. Even canning ground meat browned with onions and seasonings can make it easier to make soup, a casserole, or pasta sauce in no time.

- For instance, if you can 12 quarts of ground beef with onions, garlic, salt, and pepper, that ground beef can be used on pizzas and in spaghetti sauce, sloppy joes, tacos, burritos, casseroles, potpies, or anything else your family is in the mood to eat.

- Another fun and rewarding way to approach canning prepared foods and meal starters is to assess what fresh foods you have on hand and in abundance. This is especially helpful when you have a home garden, find a great sale, or are gifted with buckets of blueberries from a gardening neighbor. In most cases, you would try to eat as much as possible and end up throwing or giving away the rest. Instead, look at the food and ask yourself what you would do with it if you had unlimited time. Would you make pies and cobblers out of that bucket of blueberries? Then can a batch of blueberry pie filling that will make you dozens of pies and cobblers when you're ready to bake them. Do you have a lot of tomatoes from your own garden? Spend a day or two making them into spaghetti sauce, marinara sauce, and tomato soup, and you won't have to worry about waste.

- Nothing is quite as satisfying as knowing that you can put a meal on the table in a flash, regardless of how busy you are, how tight money may be, or how many guests pop in for a visit. Canning prepared foods and meal starters gives you that satisfaction.

Ground Meat Base Mix

This basic mixture is a great foundation for any number of meals, from soups and stews to casseroles and tacos. Having it already cooked and on hand will make so many mealtimes easier and less stressful.

- 8 pounds ground beef, lamb, pork, chicken, or turkey
- 5 teaspoons salt
- 2 teaspoons freshly ground black pepper
- 2 medium yellow onions, chopped
- 4 large garlic cloves, chopped
- 3 cups beef or chicken broth
- 7 quart jars, lids, and bands

1. Prepare your jars by bringing them to a boil in a large pot. Once they reach a boil, reduce the heat to low, and allow them to simmer until you're ready to use them. Prepare the lids and bands by simmering (not boiling) them over low heat in a small saucepan.

2. Prepare your canner by filling it with 2 to 3 inches of water and bringing it to a boil over high heat.

3. In a large bowl, combine the meat, salt, and pepper, mixing well with clean hands. Working in batches, if necessary, brown the meat over medium-heat in a large skillet. Stir frequently with a spatula or wooden spoon, breaking up the meat as it cooks to a crumbly texture.

4. Once the meat is gently browned, add the onions and garlic and sauté for another 3 to 4 minutes or until the onions are transparent. Reduce the heat to low.

5. In a large saucepan, heat the broth over medium-high heat until it boils and then continue a soft boil.

6. Pack the hot meat loosely into the jars, leaving 1 inch of headspace. Add enough hot broth to cover, leaving 1 inch of headspace.

7. Remove air bubbles, wipe rims, center the lids, and screw on the bands and adjust until they are fingertip tight. Place the jars in the canner.

8. Cover, vent, and pressurize the canner according to its manufacturer's directions.

9. Process for 90 minutes at 11 pounds of pressure, adjusting for altitude. Remove the canner from the burner, and follow the directions for handling noted in "Basic Instructions for Pressure Canning."

Chicken with Gravy

You can make a lot of great meals starting with just cooked chicken and a little pan gravy. Use this as the starter for chicken potpie, chicken and dumplings, chicken noodle casserole, or anything else you and your family enjoy. This recipe works equally well for turkey.

- 10 pounds bone-in chicken breasts or thighs
- 3 tablespoons olive or canola oil
- 4 teaspoons salt
- 2 teaspoons freshly ground black pepper
- 8 tablespoons reserved chicken drippings
- 8 tablespoons flour
- 6 cups chicken broth or stock
- 1 teaspoon dried rosemary
- 1 teaspoon dried thyme
- 7 quart jars, lids, and bands

1. Prepare your jars by bringing them to a boil in a large pot. Once they reach a boil, reduce the heat to low, and allow them to simmer until you're ready to use them. Prepare the lids and bands by simmering (not boiling) them over low heat in a small saucepan.

2. Prepare your canner by filling it with 2 to 3 inches of water and bringing it to a boil over high heat.

3. Preheat the oven to 375 degrees F.

4. Brush the chicken with the oil, and then season on all sides with the salt and pepper. Place into one or two large baking dishes, and bake for 40 to 45 minutes or until the chicken is cooked through. Set aside in the casserole dishes for 10 to 15 minutes or until cool enough to handle.

5. Remove all of the meat and skin from the chicken. Reserve bones and scraps for stock if desired.

6. Measure out 8 tablespoons of chicken drippings from the pans and place in a large saucepan. Bring to a boil over medium-high heat.

7. Once the drippings boil, quickly add the flour, whisking constantly until it resembles a paste.

8. Slowly stir in the broth or stock, and whisk until smooth. Add the rosemary and thyme. Bring back to a boil and then simmer.

9. Pack the meat loosely into the jars, leaving 2 inches of headspace. Add enough gravy to cover and leave 1 inch of headspace.

10. Remove air bubbles, wipe rims, center the lids, and screw on the bands and adjust until they are fingertip tight. Place the jars in the canner.

11. Cover, vent, and pressurize the canner according to its manufacturer's directions.

12. Process for 90 minutes at 11 pounds of pressure, adjusting for altitude. Remove the canner from the burner, and follow the directions for handling noted in "Basic Instructions for Pressure Canning."

Meatless Spaghetti Sauce

This basic spaghetti sauce is versatile enough to use as a pizza sauce, for pasta dishes, or on sub sandwiches. It's quick and easy to make, and it tastes wonderful.

- 30 pounds fresh tomatoes
- ¼ cup vegetable oil
- 1 cup chopped yellow onion
- 5 large garlic cloves, chopped
- 1 cup chopped green bell pepper
- 1 pound sliced fresh mushrooms
- 4½ teaspoons salt
- 2 teaspoons freshly ground black pepper
- 4 tablespoons dried parsley
- 2 tablespoons dried oregano
- ¼ cup packed brown sugar
- 9 pint jars, lids, and bands

1. Prepare your jars by bringing them to a boil in a large pot. Once they reach a boil, reduce the heat to low, and allow them to simmer until you're ready to use them. Prepare lids and bands by simmering (not boiling) them over low heat in a small saucepan.

2. Prepare your canner by filling it with 2 to 3 inches of water and bringing it to a boil over high heat.

3. Bring a large pot of water to a boil over high heat.

4. Wash the tomatoes and plunge into the boiling water for 30 seconds or until the skins begin to split. Dip immediately into icy cold water, and slip off the skins. Core and quarter the tomatoes.

5. In a large saucepan, bring the tomatoes to a boil over medium-high heat, stirring frequently. Boil for 20 minutes, uncovered. Put through a food mill or sieve if you would like the sauce to be smoother.

6. In a large saucepan, heat the vegetable oil over medium-high heat and sauté the onion, garlic, bell pepper, and mushrooms until tender. Combine the sautéed vegetables and the tomatoes, and add the salt, black pepper, and herbs. Once the sauce boils, stir in the brown sugar. Simmer uncovered, until reduced by about half.

7. Ladle the hot spaghetti sauce into the jars, leaving 1 inch of headspace.

8. Remove air bubbles, wipe rims, center the lids, and screw on the bands and adjust until they are fingertip tight. Place the jars in the canner.

9. Cover, vent, and pressurize the canner according to its manufacturer's directions.

10. Process for 20 minutes at 11 pounds of pressure, adjusting for altitude. Remove the canner from the burner, and follow the directions for handling noted in "Basic Instructions for Pressure Canning."

Spaghetti Sauce with Meat

This sauce is a hearty and very flavorful one that will serve you well for spaghetti, lasagna, or sub sandwiches.

- 30 pounds fresh tomatoes
- ¼ cup chopped fresh parsley
- ¼ cup chopped fresh basil
- 4 tablespoons chopped fresh oregano
- ¼ cup vegetable oil
- 5 pounds fresh ground beef, chicken, or turkey
- 4½ teaspoons salt
- 2 teaspoons freshly ground black pepper
- 5 large garlic cloves, chopped
- 1 cup chopped green bell pepper
- 1 cup chopped yellow onion
- 1 pound sliced fresh mushrooms
- 2 tablespoons dark brown sugar
- 7 quart jars, lids, and bands

1. Prepare your jars by bringing them to a boil in a large pot. Once they reach a boil, reduce the heat to low, and allow them to simmer until you're ready to use them. Prepare the lids and bands by simmering (not boiling) them over low heat in a small saucepan.

2. Prepare your canner by filling it with 2 to 3 inches of water and bringing it to a boil over high heat.

3. Bring a large pot of water to a boil over high heat.

4. Wash the tomatoes and plunge them into the boiling water for 30 seconds or until the skins begin to split. Dip the tomatoes immediately into icy cold water and slip off the skins. Core and quarter the tomatoes.

5. In a large saucepan, bring the tomatoes to a boil over medium-high heat, stirring frequently. Add the parsley, basil, and oregano. Boil for 20 minutes, uncovered. Put through a food mill or sieve if you would like the sauce to be smoother.

6. Meanwhile, in a large skillet, heat the vegetable oil over medium-high heat and add the ground meat, salt, black pepper, garlic, bell pepper, and onion. Sauté until lightly browned and then stir in the mushrooms. Continue cooking until the meat is cooked through.

7. Combine the sautéed meat and the tomatoes, and return to a boil. Stir in the brown sugar and simmer uncovered, until reduced by about half.

8. Ladle the hot spaghetti sauce into the jars, leaving 1 inch of headspace.

9. Remove air bubbles, wipe rims, center the lids, and screw on the bands and adjust until they are fingertip tight. Place the jars in the canner.

10. Cover, vent, and pressurize the canner according to its manufacturer's directions.

11. Process for 90 minutes at 11 pounds of pressure, adjusting for altitude. Remove the canner from the burner, and follow the directions for handling noted in "Basic Instructions for Pressure Canning."

Vegetable Soup

This vegetable soup is a great one for using plenty of fresh vegetables when they're abundant in your garden or at their peak and priced low at the farmers' market.

- 8 cups peeled and chopped tomatoes
- 6 cups peeled and diced potatoes
- 6 cups sliced carrots
- 4 cups green beans, trimmed and cut into 1-inch pieces
- 4 cups uncooked corn kernels
- 2 cups celery, sliced 1-inch thick

- 2 cups chopped yellow onions
- 6 cups water
- 2 teaspoons salt
- 1 teaspoon freshly ground black pepper
- ½ cup chopped fresh parsley
- 1 tablespoon chopped fresh rosemary
- 7 quart jars, lids, and bands

1. Prepare your jars by bringing them to a boil in a large pot. Once they reach a boil, reduce the heat to low, and allow them to simmer until you're ready to use them. Prepare the lids and bands by simmering (not boiling) them over low heat in a small saucepan.

2. Prepare your canner by filling it with 2 to 3 inches of water and bringing it to a boil over high heat.

3. Combine all of the vegetables in a large stockpot over medium-high heat. Add the water. Season with salt and pepper, and stir well. Add in the herbs and stir.

4. Bring to a boil. Reduce the heat and simmer 25 minutes.

5. Ladle the hot soup into the jars, leaving 1 inch of headspace.

6. Remove air bubbles, wipe rims, center the lids, and screw on the bands and adjust until they are fingertip tight. Place the jars in the canner.

7. Cover, vent, and pressurize the canner according to its manufacturer's directions.

8. Process for 1 hour and 25 minutes at 10 pounds of pressure, adjusting for altitude. Remove the canner from the burner, and follow the directions for handling noted in "Basic Instructions for Pressure Canning."

Tomato Soup

This soup is loaded with flavor and so much better than what you can buy in the store. There's nothing better with a grilled cheese sandwich.

- 8 quarts cored, peeled, and chopped ripe tomatoes
- 6 cups chopped onions
- 4 cups chopped celery
- 4 cups chopped green bell peppers
- 3 cups sliced carrots
- 4 teaspoons salt
- 4 teaspoons chopped fresh basil
- 2 bay leaves
- 12 pint jars, lids, and bands

1. Prepare your jars by bringing them to a boil in a large pot. Once they reach a boil, reduce the heat to low, and allow them to simmer until you're ready to use them. Prepare the lids and bands by simmering (not boiling) them over low heat in a small saucepan.

2. Prepare your canner by filling it with 2 to 3 inches of water and bringing it to a boil over high heat.

3. In a large stockpot over medium-high heat, simmer the tomatoes for 20 minutes or until soft. Press through a food mill, or process until smooth in a food processor.

4. In a large stockpot over high heat, combine all of the vegetables, salt, basil, and bay leaves and add just enough water to cover. Bring to a boil, reduce the heat to medium, and simmer for 30 minutes or until the vegetables are tender. Drain, remove the bay leaves, and press through a food mill or process until smooth in a food processor.

5. Combine the tomato puree and the veggie puree in a large stockpot and bring to a boil. Reduce the heat and simmer until thickened, stirring frequently.

6. Ladle the hot soup into the jars, leaving ½ inch of headspace.

7. Remove air bubbles, wipe rims, center the lids, and screw on the bands and adjust until they are fingertip tight. Place the jars in the canner.

8. Cover, vent, and pressurize the canner according to its manufacturer's directions.

9. Process for 20 minutes at 10 pounds of pressure, adjusting for altitude. Remove the canner from the burner, and follow the directions for handling noted in "Basic Instructions for Pressure Canning."

Chicken Soup

This hearty chicken soup is a classic and a real favorite among home canners. This makes a wonderful gift for a new mother or someone who is ill. To make it even heartier, boil some egg noodles as you're reheating your soup and stir them into the soup at the end.

- 16 cups chicken stock
- 3 cups diced cooked chicken
- 1½ cups diced celery
- 1½ cups sliced carrots
- 1 cup diced onion
- 2 teaspoons salt
- 1 teaspoon freshly ground black pepper
- ½ cup chopped fresh parsley
- 2 tablespoons chopped fresh thyme
- 1½ teaspoons turmeric
- Chicken bouillon cubes (optional)
- 4 quart jars, lids, and bands

1. Prepare your jars by bringing them to a boil in a large pot. Once they reach a boil, reduce the heat to low, and allow them to simmer until you're ready to use them. Prepare the lids and bands by simmering (not boiling) them over low heat in a small saucepan.

2. Prepare your canner by filling it with 2 to 3 inches of water and bringing it to a boil over high heat.

3. In a large stockpot over medium-high heat, combine the chicken stock, chicken, celery, carrots, and onion. Bring to a boil, reduce the heat to medium, and add the salt, pepper, parsley, thyme, and turmeric, stirring well.

4. Cover and simmer for 30 minutes. Add bouillon cubes, if desired. Cook until the bouillon cubes are dissolved.

5. Ladle the hot soup into the jars, leaving 1 inch of headspace.

6. Remove air bubbles, wipe rims, center the lids, and screw on the bands and adjust until they are fingertip tight. Place the jars in the canner.

7. Cover, vent, and pressurize the canner according to its manufacturer's directions.

8. Process for 90 minutes at 10 pounds of pressure, adjusting for altitude. Remove the canner from the burner, and follow the directions for handling noted in "Basic Instructions for Pressure Canning."

Mexican-Style Chicken Soup

When you're in the mood for a zesty, slightly spicy bowl of soup, this is the one to try. All you'll have to do for a great meal is heat some tortillas and throw together a salad.

- 4 pints home-canned tomatoes
- 4 pints diced tomatoes (fresh or canned)
- 6 cups water
- 6 cups chicken broth
- 2 (15-ounce) cans black beans, rinsed and drained
- 4 cups sliced celery
- 3 cups fresh-cut corn kernels
- 3 cups sliced carrots
- 2 large red onions, chopped
- 2 tablespoons salt
- 2 teaspoons ground cumin
- 6 large cooked chicken breasts, shredded
- ¼ cup chopped fresh cilantro
- 7 quart jars, lids, and bands

1. Prepare your jars by bringing them to a boil in a large pot. Once they reach a boil, reduce the heat to low, and allow them to simmer until you're ready to use them. Prepare the lids and bands by simmering (not boiling) them over low heat in a small saucepan.

2. Prepare your canner by filling it with 2 to 3 inches of water and bringing it to a boil over high heat.

3. In a large stockpot, combine all of the ingredients except the chicken and cilantro. Bring to a boil, cover, and simmer 3 minutes. Add the chicken and cilantro, and boil 5 minutes longer.

4. Ladle the hot soup into the jars, leaving 1 inch of headspace.

5. Remove air bubbles, wipe rims, center the lids, and screw on the bands and adjust until they are fingertip tight. Place the jars in the canner.

6. Cover, vent, and pressurize the canner according to its manufacturer's directions.

7. Process for 90 minutes at 11 pounds of pressure, adjusting for altitude. Remove the canner from the burner, and follow the directions for handling noted in "Basic Instructions for Pressure Canning."

Chunky Beef Stew

Chunky Beef Stew

Beef stew is a great prepared meal to can at home. You can mix and match various vegetables depending on what you have on hand, and everyone loves a hot bowl of stew with some biscuits.

- 1 tablespoon vegetable oil
- 5 pounds beef stew meat, cut into 1½-inch cubes
- 12 cups peeled and cubed new potatoes
- 8 cups sliced carrots
- 3 cups chopped celery
- 3 cups chopped onions
- 1½ tablespoons salt
- 1 teaspoon dried thyme
- 1 teaspoon freshly ground black pepper
- 2 bay leaves
- 7 quart jars, lids, and bands

1. Prepare your jars by bringing them to a boil in a large pot. Once they reach a boil, reduce the heat to low, and allow them to simmer until you're ready to use them. Prepare the lids and bands by simmering (not boiling) them over low heat in a small saucepan.

2. Prepare your canner by filling it with 2 to 3 inches of water and bringing it to a boil over high heat.

3. Fill another large pot with water and bring to a boil. Continue boiling until ready to use.

4. In a very large, heavy skillet, heat the oil over medium-high heat and brown the meat on all sides, stirring often.

5. Add the vegetables and seasonings, and then add enough boiling water to cover. Cook for another 5 minutes, stirring often.

6. Ladle the hot stew into the jars, leaving 1 inch of headspace.

7. Remove air bubbles, wipe rims, center the lids, and screw on the bands and adjust until they are fingertip tight. Place the jars in the canner.

8. Cover, vent, and pressurize the canner according to its manufacturer's directions.

9. Process for 90 minutes at 10 pounds of pressure, adjusting for altitude. Remove the canner from the burner, and follow the directions for handling noted in "Basic Instructions for Pressure Canning."

Hearty Chili

People don't just like chili; they love it. Fortunately, it's a filling and economical meal that's great for lunch or dinner in any season. This chili is especially hearty and will be a big hit with your chili fans.

- 4 pounds boneless beef chuck roast or steak
- ¼ cup vegetable oil
- 3 cups chopped yellow onions
- 2 large garlic cloves, chopped
- 5 tablespoons chili powder
- 2 teaspoons cumin seed
- 2 teaspoons salt
- 1 teaspoon dried oregano
- ½ teaspoon freshly ground black pepper
- ½ teaspoon coriander
- ½ teaspoon crushed red pepper
- 6 cups diced canned tomatoes with their juice
- 4 quart jars, lids, and bands

1. Prepare your jars by bringing them to a boil in a large pot. Once they reach a boil, reduce the heat to low, and allow them to simmer until you're ready to use them. Prepare the lids and bands by simmering (not boiling) them over low heat in a small saucepan.

2. Prepare your canner by filling it with 2 to 3 inches of water and bringing it to a boil over high heat.

3. Cut the beef chuck into ½-inch cubes, removing any sinew or extra fat.

4. Heat the oil in a large, heavy stockpot and lightly brown the meat. Add the onions and garlic, and continue to cook until the onions are slightly soft.

5. Add all of the seasoning and spices, and cook for 5 minutes. Stir in the diced tomatoes, together with their juices.

6. Bring to a boil, reduce the heat to medium, and simmer for 50 minutes, stirring occasionally.

7. Ladle the hot chili into the jars, leaving 1 inch of headspace.

8. Remove air bubbles, wipe rims, center the lids, and screw on the bands and adjust until they are fingertip tight. Place the jars in the canner.

9. Cover, vent, and pressurize the canner according to its manufacturer's directions.

10. Process for 90 minutes at 10 pounds of pressure, adjusting for altitude. Remove the canner from the burner, and follow the directions for handling noted in "Basic Instructions for Pressure Canning."

Chili con Carne

This chili recipe is a pretty traditional take on chili and will appeal to those who prefer a ground beef chili with beans. You can also make this chili with ground pork or venison.

- 3 cups dried pinto or red kidney beans
- 5½ cups water
- 5 teaspoons salt, divided
- 3 pounds ground beef
- 1½ cups chopped onion
- 1 cup chopped hot peppers of your choice (optional)
- 1 teaspoon freshly ground black pepper
- 3–6 tablespoons chili powder
- 2 quarts partly crushed whole tomatoes
- 9 pint jars, lids, and bands

---

1. Wash the kidney beans thoroughly and place them in a large saucepan. Add enough cold water to cover 2 to 3 inches above the beans and soak 12 to 18 hours.

2. Drain the beans thoroughly.

3. Prepare your jars by bringing them to a boil in a large pot. Once they reach a boil, reduce the heat to low, and allow them to simmer until you're ready to use them. Prepare the lids and bands by simmering (not boiling) them over low heat in a small saucepan.

4. Prepare your canner by filling it with 2 to 3 inches of water and bringing it to a boil over high heat.

5. In a large stockpot, combine the beans with the 5½ cups of water and 2 teaspoons of the salt. Bring to a boil, reduce the heat to medium, and simmer for 30 minutes. Drain and discard the water.

6. In a heavy skillet over medium-high heat, brown the ground beef, onion, and hot peppers.

7. Drain off the fat and add the remaining salt, black pepper, chili powder, and tomatoes. Add the drained beans. Simmer 5 minutes. Do not thicken the liquid!

8. Ladle the hot chili into the jars, leaving 1 inch of headspace.

9. Remove air bubbles, wipe rims, center the lids, and screw on the bands and adjust until they are fingertip tight. Place the jars in the canner.

10. Cover, vent, and pressurize the canner according to its manufacturer's directions.

11. Process for 75 minutes at 11 pounds of pressure, adjusting for altitude. Remove the canner from the burner, and follow the directions for handling noted in "Basic Instructions for Pressure Canning."

Apple Pie Filling

Almost everyone loves apple pie, but not too many of us are eager to peel all of those apples every week. With your pressure canner and this recipe, you can peel them once and have enough apple pie filling to make ten or twelve pies. This is great for those times when you need to bring a dish to a potluck or barbecue.

- 4 cups granulated sugar
- 1 cup cornstarch
- 1 tablespoon ground cinnamon
- 1 teaspoon ground cardamom
- 1 teaspoon ground nutmeg
- 9 cups water
- ¼ cup lemon juice
- 2 teaspoons lemon zest
- 9 pounds peeled, cored, and sliced apples
- 6 quart jars, lids, and bands

1. Prepare your jars by bringing them to a boil in a large pot. Once they reach a boil, reduce the heat to low, and allow them to simmer until you're ready to use them. Prepare the lids and bands by simmering (not boiling) them over low heat in a small saucepan.

2. Prepare your canner by filling it with 2 to 3 inches of water and bringing it to a boil over high heat.

3. In a large stockpot over medium-high heat, combine the sugar, cornstarch, spices, and water and bring to a boil, stirring constantly. Boil until thick and bubbly.

4. Reduce the heat to medium, add the lemon juice, lemon zest, and apples and cook for 5 minutes, stirring constantly.

5. Ladle the hot filling into the jars, leaving 1 inch of headspace.

6. Remove air bubbles, wipe rims, center the lids, and screw on the bands and adjust until they are fingertip tight. Place the jars in the canner.

7. Cover, vent, and pressurize the canner according to its manufacturer's directions.

8. Process for 8 minutes at 6 pounds of pressure, adjusting for altitude. Remove the canner from the burner, and follow the directions for handling noted in "Basic Instructions for Pressure Canning."

Blueberry Pie Filling

Blueberries have a short growing season and tend to ripen all at once. If you have one blueberry bush or three, you may find yourself with loads of berries that need to be eaten right away. (If not, grab them in July when everyone else is overloaded and prices are low!) Nothing celebrates blueberries like a fresh pie, so make several batches of this delicious pie filling to enjoy all year long.

- 10 pints blueberries
- ½ cup white or purple grape juice
- ½ cup granulated sugar
- 6 tablespoons Clear Jel (available where canning supplies are sold)

- 4 tablespoons lemon juice
- 2 teaspoons almond extract (optional)
- 2 quart jars, lids, and bands

---

1. Stem, wash, and sort your berries very well, and allow to drain on paper towels.

2. Prepare your jars by bringing them to a boil in a large pot. Once they reach a boil, reduce the heat to low, and allow them to simmer until you're ready to use them. Prepare the lids and bands by simmering (not boiling) them over low heat in a small saucepan.

3. Prepare your canner by filling it with 2 to 3 inches of water and bringing it to a boil over high heat.

4. Combine the grape juice, sugar, Clear Jel, and lemon juice in a heavy pot over medium heat, and whisk constantly for about 2 minutes until thickened. Continue to whisk for 1 minute more.

5. Add the blueberries and stir well, and then continue cooking, stirring often, until the mixture returns to a boil. Boil for 5 minutes. Stir in the almond extract, if desired, and remove from heat.

6. Ladle the hot filling into the jars, leaving ½ inch of headspace.

7. Remove air bubbles, wipe rims, center the lids, and screw on the bands and adjust until they are fingertip tight. Place the jars in the canner.

8. Cover, vent, and pressurize the canner according to its manufacturer's directions.

9. Process for 8 minutes at 6 pounds of pressure, adjusting for altitude. Remove the canner from the burner, and follow the directions for handling noted in "Basic Instructions for Pressure Canning."

Variations:

This recipe works beautifully for Gooseberry Pie Filling, Raspberry Pie Filling, and Blackberry Pie Filling. Omit the almond extract for the gooseberries, and substitute 1 teaspoon of lemon zest.

Peach Pie Filling

Peach pie can be a lot of work when you're making it from scratch, but the difference in flavor is enormous. The way to enjoy that flavor is to do all of the work in one day; then enjoy pies all year round. If you have a peach tree or know someone who does, this is one of the best ways to use the bounty.

- 6 quarts peeled and sliced fresh peaches
- Citric acid solution (see below)
- 5¼ cups water
- 7 cups granulated sugar
- 2 cups plus 3 tablespoons Clear Jel, divided (available where canning supplies are sold)
- 1 teaspoon almond extract
- 1¼ cups lemon juice
- 7 quart jars, lids, and bands

1. Prepare your jars by bringing them to a boil in a large pot. Once they reach a boil, reduce the heat to low, and allow them to simmer until you're ready to use them. Prepare the lids and bands by simmering (not boiling) them over low heat in a small saucepan.

2. Prepare your canner by filling it with 2 to 3 inches of water and bringing it to a boil over high heat.

3. Place the sliced peaches in a citric acid solution. If you use citric acid, 1 level teaspoon is needed per gallon of water. If using vitamin C tablets, use 6 tablets per gallon of water, crushing them before adding them to the water. Allow the peaches to remain in the solution for at least 5 minutes.

4. In a large stockpot, bring 1 gallon of water to a boil. Boil the peaches 6 cups at a time for 1 minute after the water returns to a boil. Drain each batch and keep covered in a heavy pot over the lowest heat to keep warm.

5. Combine the water, sugar, Clear Jel, and almond extract in a large stockpot over medium-high heat. Stir and cook until the mixture thickens and starts to bubble. Add the lemon juice and boil for 1 minute more, stirring constantly.

6. Fold in the drained peach slices and continue to cook for 3 minutes.

7. Pack the hot peaches into the jars, leaving 1 inch of headspace.

8. Remove air bubbles, wipe rims, center the lids, and screw on the bands and adjust until they are fingertip tight. Place the jars in the canner.

9. Cover, vent, and pressurize the canner according to its manufacturer's directions.

10. Process for 8 minutes at 6 pounds of pressure, adjusting for altitude. Remove the canner from the burner, and follow the directions for handling noted in "Basic Instructions for Pressure Canning."

CONCLUSION

Once you get the hang of canning (and that won't take long at all), a whole world of possibilities opens up to you. You'll find yourself able to home-can almost any foods. You'll also discover that doing so can have a wonderful impact on your home, life, and budget.

Canning as a method of preserving food has shown a great resurgence in recent years. As this book has shown, it's a way to live less expensively, more sustainably, and more independently. People with pantries full of home-canned (and even homegrown) foods are not at the total mercy of grocery stores and their price markups. Canning gives you a great deal of independence, even if you're canning only small batches of a few foods. It allows you to buy more locally and more seasonally, and enjoy those foods all year round, long after they're out of season.

Canning may be hard work when you're in the midst of it. It may involve a lot of chopping, boiling, and packing. But once the pots are washed and the jars have cooled, nothing is quite as satisfying as surveying shelf after shelf loaded with jewel-colored jars. That satisfaction can be as simple as knowing that you accomplished something or as meaningful as knowing that if you are faced with tough economical times, you have enough high-quality food to see your family through.

Canning is one of the best ways to share good food with your family, your friends, and your community. Once you get started, you're likely to get hooked.

ALTITUDE CHART

Below are altitudes of selected cities in the United States and Canada. To find the exact altitude of your location, use the search features on the EarthTools website (www.earthtools.org).

STATE	CITY	FEET	METERS
ARIZONA	Mesa	1,243	379
	Phoenix	1,150	351
	Scottsdale	1,257	383
	Tucson	2,389	728
CALIFORNIA	Fontana	1,237	377
	Moreno Valley	1,631	497
COLORADO	Aurora	5,471	1,668
	Colorado Springs	6,010	1,832
	Denver	5,183	1,580
GEORGIA	Atlanta	1,026	313
IDAHO	Boise	2,730	832
	Idaho Falls	4,705	1,434
IOWA	Sioux City	1,201	366
KANSAS	Wichita	1,299	396
MONTANA	Billings	3,123	952
	Missoula	3,209	978
NEBRASKA	Lincoln	1,176	358
	Omaha	1,090	332
NEVADA	Henderson	1,867	569
	Las Vegas	2,001	610
	Reno	4,505	1,373
NEW MEXICO	Albuquerque	5,312	1,619
	Santa Fe	7,260	2,213
NORTH CAROLINA	Asheville	2,134	650
NORTH DAKOTA	Bismarck	1,686	514
OHIO	Akron	1,004	306
OKLAHOMA	Oklahoma City	1,201	366
PENNSYLVANIA	Pittsburgh	1,370	418
SOUTH DAKOTA	Rapid City	3,202	976
TEXAS	Amarillo	3,605	1,099
	El Paso	3,740	1,140
	Lubbock	3,256	992
UTAH	Provo	4,551	1,387
	Salt Lake City	4,226	1,288
WASHINGTON	Spokane	1,843	562
WYOMING	Casper	5,150	1,570

PROVINCE	CITY	FEET	METERS
ALBERTA	Calgary	3,600	1,100
	Edmonton	2,201	671
ONTARIO	Hamilton	1,063	324
MANITOBA	Brandon	1,343	409
SASKATCHEWAN	Regina	1,893	577
	Saskatoon	1,580	482

GLOSSARY

alum   a compound that was widely used in home canning to keep pickles, cucumbers, and some other fresh foods crisp. It's rarely used anymore, but you may see it in some heirloom recipes.

boiling water canning   see water bath canning

citric acid solution   a solution used in canning to prevent darkening of foods such as potatoes, apples, pears, and peaches. Can be made from citric acid powder or by using vitamin C tablets.

dial-gauge pressure canner   this pressure canner uses a dial to indicate when the proper pressure has been reached within the pressure canner.

high-acid foods   foods with enough acid to make them safe for water bath canning. This includes foods with a pH of 4.6 or lower.

jam   smooth, spreadable fruit product made with pureed or crushed fruit.

jelly   a clear, spreadable fruit product made from only the juice of the fruit.

low-acid foods   foods that lack the acid content to be safely canned in a water bath canner. These include foods with a pH of 4.6 or higher.

pectin   a naturally occurring compound that is present in the skin and pulp of many fruits, especially apples. It is used to thicken the liquid in jams and jellies.

pH   the measurement of the amount of acid in foods.

preserves   a spreadable fruit product made with whole or mostly whole fruit. It is thicker than jams.

pressure canning   the process of canning using both heat and pressure. Essentially, food is packed into jars and steamed within the canner. It is always used for canning low-acid foods, but can also be used to can high-acid foods.

processing time   the length of time a food must be processed in a water bath canner or pressure canner. In a water bath canner, that time starts from the moment the water returns to a boil. In a pressure canner, it begins from the time the canner has reached the proper pressure.

psi   pounds per square inch. It is used in pressure canning to indicate the proper pressure for processing each food.

water bath canning   the process of canning using boiling water. It can be used only for high-acid foods.

weighted-gauge pressure canner   most older models of pressure canners and some newer models are weighted-gauge canners, with a weight attached to the vent of the canner that jiggles and rattles when the proper pressure is reached and maintained.

INDEX

Made in the USA
Middletown, DE
13 November 2018